Elementary Euclidean Geometry
An Introduction

This is a genuine introduction to the geometry of lines and conics in the Euclidean plane. Lines and circles provide the starting point, with the classical invariants of general conics introduced at an early stage, yielding a broad subdivision into types, a prelude to the congruence classification. A recurring theme is the way in which lines intersect conics. From single lines one proceeds to parallel pencils, leading to midpoint loci, axes and asymptotic directions. Likewise, intersections with general pencils of lines lead to the central concepts of tangent, normal, pole and polar.

The treatment is example-based and self-contained, assuming only a basic grounding in linear algebra. With numerous illustrations and several hundred worked examples and exercises, this book is ideal for use with undergraduate courses in mathematics, or for postgraduates in engineering and the physical sciences.

C. G. GIBSON is a senior fellow in mathematical sciences at the University of Liverpool.

Elementary Euclidean Geometry

An Introduction

C. G. GIBSON

WB

PUBLISHED BY THE PRESS SYNDICATE OF THE UNIVERSITY OF CAMBRIDGE
The Pitt Building, Trumpington Street, Cambridge, United Kingdom

CAMBRIDGE UNIVERSITY PRESS
The Edinburgh Building, Cambridge CB2 2RU, UK
40 West 20th Street, New York, NY 10011–4211, USA
477 Williamstown Road, Port Melbourne, VIC 3207, Australia
Ruiz de Alarcón 13, 28014 Madrid, Spain
Dock House, The Waterfront, Cape Town 8001, South Africa

http://www.cambridge.org

First published 2003

Printed in the United Kingdom at the University Press, Cambridge

Typeface Times 10/13 pt.   *System* LATEX 2_ε [TB]

A catalogue record for this book is available from the British Library

Library of Congress Cataloguing in Publication data

Gibson, Christopher G., 1940–
Elementary Euclidean geometry: an introduction / C.G. Gibson.
p.   cm.
Includes bibliographical references and index.
ISBN 0 521 83448 1
1. Geometry.   I. Title.

QA453.G45 2004 516.2–dc22   2003055904

ISBN 0 521 83448 1 hardback

1.1 / 2 / 04

Contents

Figures

Tables

Preface

It is worth saying something about the background to this book, since it is linked to a sea change in the teaching of university mathematics, namely the renaissance in undergraduate geometry, following a postwar decline. There is little doubt that the enormous progress made in studying non-linear phenomena by geometric methods has rekindled interest in the subject. However, that is not the only reason for seeking change, as I pointed out in the preface to *Elementary Geometry of Algebraic Curves*:

'For some time I have felt there is a good case for raising the profile of undergraduate geometry. The case can be argued on *academic* grounds alone. Geometry represents a way of thinking within mathematics, quite distinct from algebra and analysis, and so offers a fresh perspective on the subject. It can also be argued on purely *practical* grounds. My experience is that there is a measure of concern in various practical disciplines where geometry plays a substantial role (engineering science for instance) that their students no longer receive a basic geometric training. And thirdly, it can be argued on *psychological* grounds. Few would deny that substantial areas of mathematics fail to excite student interest: yet there are many students attracted to geometry by its sheer visual content.'

Background

A good starting point in developing undergraduate geometry is to focus on plane curves. They comprise a rich area, of historical significance and increasing relevance in the physical and engineering sciences. That raises a practical consideration, namely that there is a dearth of suitable course texts: some are out of date, whilst others are written at too high a level, or contain too much material.

I felt it was time to improve the situation, bearing in mind the importance of foundational mathematical training, where the primary objective is to enable students to gain fluency in the basics. (Those who wish to develop their interests will be warmly welcomed at the postgraduate level.) Over my career, one of the healthier developments in the teaching of university mathematics is the widespread adoption of clean, careful treatments of foundational material. For instance linear algebra, group theory, general abstract algebra, introductory calculus and real analysis are now widely taught on this pattern, supported by excellent texts. Such courses fit the contemporary mould of good mathematics education, by exhibiting internal coherence, an intrinsic approach, and standards of proof appropriate to the subject. I wanted to see geometry regain its place in the mathematics curriculum, within this broad pattern.

The Elementary Geometry Trilogy

It was against this background that I wrote two companion texts[1] presenting elementary accounts of complementary viewpoints, to wit the *algebraic* viewpoint (where curves are defined by the vanishing of a polynomial in two variables) and the *differentiable* viewpoint (where curves are parametrized by a single real variable). I have been encouraged by the reactions of the mathematical community, which has welcomed these contributions to undergraduate geometry.

Both texts were intended primarily for second year students, with later material aimed at third years. However, neither addresses the question of introducing university students to geometry *for the first time*. I emphasize this for good reason, namely that geometry has largely disappeared from school mathematics. In my experience, few students acquire more than an imperfect knowledge of lines and circles before embarking on their degree studies.

I think the way forward is to offer foundational geometry courses which properly expose the body of knowledge common to both viewpoints, the basic geometry of lines and conics in the Euclidean plane. The geometry of conics is important in its own right. Conics are of considerable historical significance, largely because they arise naturally in numerous areas of the physical and engineering sciences, such as astronomy, electronics, optics, acoustics, kinematics, dynamics and architecture. Quite apart from their physical importance, conics are quite fundamental objects in mathematics itself, playing crucial roles in understanding general plane curves.

---

[1] *Elementary Geometry of Algebraic Curves* and *Elementary Geometry of Differentiable Curves*, published by Cambridge University Press, and henceforth referred to as *EGAC* and *EGDC* respectively. The present text will be designated as *EEG*.

In this respect *EEG* should be of solid practical value to students and teachers alike. On such a basis, students can develop their geometry with a degree of confidence, and a useful portfolio of down-to-earth examples. For teachers, *EEG* provides a source of carefully worked out material from which to make a selection appropriate to their objectives. Such a selection will depend on several factors, such as the attainment level of the students, the teaching time available, and the intended integration with other courses.

I make no apology for the fact that some sections overlap the material of *EGAC* and *EGDC*. On the contrary, I saw close integration as a positive advantage. This book is a convenient stepping stone to those texts, taking one further down the geometry road, and bringing more advanced treatments within reach. In this way *EEG* can be viewed as the base of a trilogy, sharing a common format. In particular, the book is unashamedly example based. The material is separated into short chapters, each revolving around a single idea. That is done for good pedagogical reasons. First, students find mathematics easier to digest when it is split into a bite–sized chunks: the overall structure becomes clearer, and the end of each chapter provides a welcome respite from the mental effort demanded by the subject. Second, by pigeon–holing the material in this way the lecturer gains flexibility in choosing course material, without damaging the overall integrity. On a smaller scale, the same philosophy is pursued within individual chapters. Each chapter is divided into a number of sections, and in turn each section is punctuated by a series of 'examples', culminating in 'exercises' designed to illustrate the material, and to give the reader plenty of opportunity to master computational techniques and gain confidence.

Axioms for Writing

The material is designed to be accessible to those with minimal mathematical preparation. Basic linear algebra is the one area where some familiarity is assumed: the material of a single semester course should suffice. And it would be an advantage for the reader to feel comfortable with the concept of an equivalence relation.

One of my guiding axioms was that the content should provide the reader with a secure foundation for further study. Though elementary, it is coherent mathematics, not just a mishmash of calculations posing as geometry. There are new ways of viewing old things, concepts to be absorbed, results to contemplate, proofs to be understood, and computational techniques to master, all of which further the student's overall mathematical development. In this respect I feel it is important for the student to recognise that although geometric intuition points one in the right direction, it is no substitute for formal proof.

To be consistent with that philosophy, it is necessary to provide intrinsic definitions and argue coherently from them.

There is something to be said for ring–fencing the content of a foundational course from the outset. In the present context, I felt there was a good case for restricting the geometry entirely to the real Euclidean plane. For instance, even with that restriction there is more than enough material from which to choose. Also, at the foundational level it may be unwise to develop too many concepts. Thus complex conics are probably best left till students feel comfortable with the mechanics of handling complex numbers. Likewise, my experience suggests it is sensible to leave the projective plane till a little later in life.

The Development

One has to maintain a careful balance between theory and practice. For instance, the initial discussion of lines emphasizes the difference between a linear function on the plane and its zero set. To the student that may seem unduly pedantic, but failure to make the distinction introduces a potential source of confusion. On the other hand, since lines are quite fundamental to the development, a whole section is devoted to the practicalities of handling them efficiently. The Euclidean structure on the plane may well be familiar from a linear algebra course: nevertheless, there is a self–contained treatment, leading to the formula for the distance from a point to a line which underlies the focal constructions of conics.

Circles provide the first examples of general conics, and of the fact that a conic may not be determined by its zero set. However, we follow the pattern for lines by showing that the zero sets of *real* circles do determine the equation, a result extended (in the final chapter) to general conics with infinite zero sets. From circles it is but a short step to general conics. The classical invariants are introduced at an early stage, yielding a first broad subdivision into types, a prelude to the later congruence classification. Despite their uninteresting geometry, degenerate conics do arise naturally in families of conics as transitional types: and for that reason, a chapter is devoted to them. Likewise, a chapter is reserved for centres, since they provide basic geometric distinctions exploited in the congruence classification.

A recurring theme in the development is the way in which lines intersect conics. From single lines we progress to parallel pencils, leading to the classical midpoint locus, and the concepts of axis and asymptotic direction. In the same vein we study pencils of lines through a point on a conic, leading to the central geometric concepts of tangent and normal. Finally, the question of how a general pencil of lines interesects a conic gives rise to the classical concepts of pole and polar, and the interesting idea of the orthoptic locus.

   This text has two distinctive features. The first is that despite its intrinsic importance to the metric geometry, the classical focal construction appears later in the development than is usual. That is quite deliberate. One reason is that it aids clarity of thought. But there is also a technical reason. I wanted a method for finding foci and directrices *independent of the congruence classification*. That not only enables the student to handle a wider range of examples, but also clarifies the uniqueness question for focal constructions, a surprising omission in most texts. Another distinctive feature is that the congruence classification is left till the end. Again, that is quite deliberate. To my way of thinking, the geometry is more interesting than the listing process, so deserves to be developed first. Also, the congruence classification is a natural resting point in the student's geometric progression. Looking back, it lends cohesion to the range of examples met in the text: and looking forward, it raises fundamental questions which are better left to final year courses.

Acknowledgements

I would like to express my profound thanks to my colleagues Bill Bruce, Alex Dimca, Wendy Hawes and Ton Marar and who very kindly took the trouble to read the typescript and make detailed comment. They removed many of the errors and inconsistencies, and made innumerable constructive suggestions for improvement. In their own way they have demonstrated their own commitment to geometry.

1

Points and Lines

Lines play a fundamental role in geometry. It is not just that they occur widely in the analysis of physical problems – the geometry of more complex curves can sometimes be better understood by the way in which they intersect lines. For some readers the material of this chapter will be familiar from linear algebra, in which case it might be best just to scan the contents and proceed to Chapter 2. Even so, you are advised to look carefully at the basic definitions. It is worth understanding the difference between a linear function and its zero set: it may seem unduly pedantic, but blurring the distinction introduces a potential source of confusion. Much of this text depends on the mechanics of handling lines efficiently and for that reason Section 1.4 is devoted to practical procedures. In Section 1.5 we consider lines from the parametric viewpoint, which will be of use later when we look at the properties of conics in more detail. Finally, we go one step further by considering pencils of lines, which will play a key role in introducing axes in Chapter 7.

1.1 The Vector Structure

Throughout this text $\mathbb{R}$ will denote the set of real numbers. For linguistic variety we will refer to real numbers as *constants* (or *scalars*).[1] We will work in the familiar real plane $\mathbb{R}^2$ of elementary geometry, whose elements $Z = (x, y)$ are called *points* (or *vectors*). Recall that we can add vectors, and multiply them by constants λ, according to the familiar rules

$$(x, y) + (x', y') = (x + x', y + y'), \qquad \lambda(x, y) = (\lambda x, \lambda y).$$

Two vectors $Z = (x, y)$, $Z' = (x', y')$ are *linearly dependent* when there exist constants λ, λ' (not both zero) for which $\lambda Z + \lambda' Z' = 0$: otherwise they are

---

[1] In this text the first occurrence of an expression is always italicized, the context defining its meaning. Now and again we also italicize expressions for emphasis.

linearly independent. Thus non-zero vectors Z, Z' are linearly dependent when each is a constant multiple of the other. By linear algebra Z, Z' are linearly independent if and only if $xy' - x'y \neq 0$: and in that case linear algebra tells us that any vector can be written uniquely in the form $\lambda Z + \lambda' Z'$ for some scalars λ, λ'.

Example 1.1 The relation of linear dependence on *non-zero* vectors is an equivalence relation on the plane (with the origin deleted) and the resulting equivalence classes are *ratios*. The ratio associated to the point (x, y) is denoted $x : y$. Provided $y \neq 0$ the ratio $x : y$ can be identified with the constant x/y, whilst the ratio $(1 : 0)$ is usually denoted ∞.

1.2 Lines and Zero Sets

Our starting point is to give a careful definition of what we mean by a line. A *linear function* in x, y is an expression $ax + by + c$, where the *coefficients* a, b, c are constants, and at least one of a, b is non-zero. Suppose we have two linear functions

$$L(x, y) = ax + by + c, \qquad L'(x, y) = a'x + b'y + c'.$$

We say that L, L' are *scalar multiples* of each other when there exists a real number $\lambda \neq 0$ with $a' = \lambda a$, $b' = \lambda b$, $c' = \lambda c$. For instance, any two of the following linear functions are scalar multiples of each other

$$x - y + 1, \qquad 2x - 2y + 2, \qquad -x + y - 1.$$

This relation on linear functions is an equivalence relation, and an equivalence class is called a *line*. Our convention is that the line associated to a linear function L is denoted by the same symbol. Associated to any linear function L is its *zero set*

$$\{(x, y) \in \mathbb{R}^2 : L(x, y) = 0\}.$$

Note that any scalar multiple of L has the same zero set, so the concept makes perfect sense for lines. Instead of saying that $P = (x, y)$ is a point in the zero set, we shall (for linguistic variety) say that P *lies on* L, or that L *passes through* P.

At this point you should pause, long enough to be sure you have absorbed the preceding definitions. A line is a linear function, up to scalar multiples: it is a quite distinct object from its zero set, a set of points in the plane. The zero set of a line is completely determined by that line. In the next section we will show that conversely, a line is completely determined by its zero set, so it may seem pedantic to separate the concepts. However the 'conics' we will meet in the Chapter 4 are not necessarily determined by their zero sets, so it is wise to get into the habit of maintaining the distinction.

1.3 Uniqueness of Equations

Though elementary, the following result is conceptually important. It is the first of a sequence of results linking two disparate notions.

Theorem 1.1 *Through any two distinct points* $P = (p_1, p_2)$, $Q = (q_1, q_2)$ *there is a unique line* $ax + by + c$.

Proof To establish this fact we seek constants a, b, c (not all zero) for which

$$ap_1 + bp_2 + c = 0, \qquad aq_1 + bq_2 + c = 0. \qquad (1.1)$$

That is a linear system of two equations in the three unknowns a, b, c with matrix

$$\begin{pmatrix} p_1 & p_2 & 1 \\ q_1 & q_2 & 1 \end{pmatrix}.$$

Since P, Q are distinct, at least one of the 2×2 minors of this matrix is non-zero. (You really ought to check this.) By linear algebra, the matrix has rank 2, hence kernel rank 1. That means that there is a *non-trivial* solution (a, b, c), and that any other solution (a', b', c') is a non-zero scalar multiple. Non-triviality means that at least one of a, b, c is non-zero: in fact, at least one of a, b is non-zero, for if $a = b = 0$ then $c \neq 0$, and our linear system of equations fails to have a solution. Thus there is a line through P, Q and any other line with that property coincides with it. $\qquad \square$

Thus *a line is determined by its zero set*, meaning that if the linear functions L, L' have the same zero sets they are scalar multiples of each other: we have only to pick two distinct points in the common zero set, and apply the above result. That justifies the time-honoured practice of referring to the *equation* $L = 0$ of a line L. Strictly, that is an abbreviation for the zero set of L, but since the zero set determines L it is not too misleading. Nevertheless, you are strongly advised to maintain a crystal-clear mental distinction between lines and their zero sets.

Example 1.2 The *slope* of a line $ax + by + c = 0$ is the ratio $-a : b$. Lines of infinite slope are *vertical* and can be written in the form $x = x_0$, whilst lines of zero slope are *horizontal* and can be written in the form $y = y_0$. For non-vertical lines the slope is identified with the constant $-a/b$. Any non-vertical line can be written $y = px + q$ for some constants p, q and has slope p: likewise, any non-horizontal line can be written $x = ry + s$ for some constants r, s and has slope $1/r$. Observe that any line can be expressed in one (or both) of these forms. It will also be convenient to refer to the ratio $-b : a$ as the *direction* of the line, and any representative of this ratio as a *direction vector*: in particular, $(-b, a)$ is a direction vector for the line.

Exercise

1.3.1   Two linear functions $a_1 x + b_1 y + c_1$, $a_2 x + b_2 y + c_2$ are such that $c_1 = a_1^2 + b_1^2$, $c_2 = a_2^2 + b_2^2$. Show that if the resulting lines coincide then $a_1 = a_2$, $b_1 = b_2$.

1.4 Practical Techniques

Much of the material in this book revolves around the sheer mechanics of handling lines. In this section we introduce a small number of practical techniques, which are well worth mastering.

Example 1.3 There is an easily remembered formula for the line through p, q of the previous example. Linear algebra (or direct substitution) tells us that a solution (a, b, c) of the equations (1.1) is given by $a = p_2 - q_2$, $b = q_1 - p_1$, $c = p_1 q_2 - p_2 q_1$. Substituting for a, b, c in $ax + by + c = 0$ we see that the equation of the line is

$$\begin{vmatrix} x & y & 1 \\ p_1 & p_2 & 1 \\ q_1 & q_2 & 1 \end{vmatrix} = 0. \tag{1.2}$$

Here is a useful application. A set of points is *collinear* when there exists one line on which all the points of the set lie. Assuming there are at least two distinct points in the set, it will be collinear if and only if every other point lies on the line joining these two. Thus to check that a given set of points is collinear we need a criterion for three points to be collinear.

Example 1.4 The condition for three distinct points $P_1 = (x_1, y_1)$, $P_2 = (x_2, y_2)$, $P_3 = (x_3, y_3)$ to be collinear is that the following relation holds. Indeed they are collinear if and only if P_1 lies on the line joining P_2, P_3 so

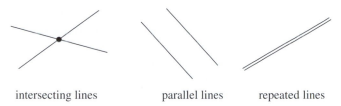

intersecting lines   parallel lines   repeated lines

Fig. 1.1. Three ways in which lines can intersect

satifies the equation of the previous example

$$\begin{vmatrix} x_1 & y_1 & 1 \\ x_2 & y_2 & 1 \\ x_3 & y_3 & 1 \end{vmatrix} = 0.$$

The *intersection* of two lines is the set of points common to both zero sets. The point of the next example is that there are just three possibilities: the intersection is either a single point, or empty, or coincides with both zero sets.

Example 1.5 The intersections of two lines L, L' are the common solutions of two linear equations

$$ax + by + c = 0, \qquad a'x + b'y + c' = 0.$$

Provided (a, b), (a', b') are linearly independent there is a unique solution, given by Cramer's Rule

$$x = \frac{bc' - b'c}{ab' - a'b}, \qquad y = \frac{a'c - ac'}{ab' - a'b}.$$

Otherwise, there are two possibilities. The first is that L, L' have no intersection, and are said to be *parallel*: and the second is that L, L' have identical zero sets, so coincide. Thus the lines parallel to $ax + by + c = 0$ are those of the form $ax + by + d = 0$ with $d \neq c$. More generally, a set of lines is *parallel* when no two of them have a common point.

A set of lines is *concurrent* when there exists a point through which every line in the set passes. Assuming there are at least two distinct lines in the set, it will be concurrent if and only if every other line passes through their intersection. It would therefore be helpful to have a criterion for three general lines to be concurrent.

Lemma 1.2 *A necessary and sufficient condition for three distinct non-parallel lines $a_1x + b_1y + c_1 = 0$, $a_2x + b_2y + c_2 = 0$, $a_3x + b_3y + c_3 = 0$ to be concurrent is that the relation (1.3) below holds*

$$\begin{vmatrix} a_1 & b_1 & c_1 \\ a_2 & b_2 & c_2 \\ a_3 & b_3 & c_3 \end{vmatrix} = 0. \tag{1.3}$$

Proof By linear algebra (1.3) is a necessary and sufficient condition for the following homogeneous sytem of linear equations to have a non-trivial solution (x, y, z)

$$a_1x + b_1y + c_1z = 0, \qquad a_2x + b_2y + c_2z = 0, \qquad a_3x + b_3y + c_3z = 0.$$

If the lines are concurrent, there is a point (p, q) lying on all three, and hence a non-trivial solution $x = p$, $y = q$, $z = 1$ of the system with $z \neq 0$. And, conversely, if there is a solution with $z \neq 0$ then the point (p, q) with $p = x/z$, $q = y/z$ lies on all three lines, so they are concurrent. It remains to consider the possibility when there is a non-trivial solution (x, y, z) with $z = 0$, so there is a non-trivial solution (x, y) for the homogeneous system

$$a_1x + b_1y = 0, \qquad a_2x + b_2y = 0, \qquad a_3x + b_3y = 0.$$

However, in that case linear algebra tells us that the vectors (a_1, b_1), (a_2, b_2), (a_3, b_3) are linearly dependent, so the lines are parallel, contrary to assumption. $\square$

Exercises

1.4.1   In each of the following cases find the equation of the line L through the given points P, Q:

      (i) $P = (1, -1)$, $Q = (2, -3)$,
      (ii) $P = (1, 7)$, $Q = (3, -4)$,
      (iii) $P = (3, -2)$, $Q = (5, -1)$.

1.4.2   In each of the following cases find the points of intersection of the given lines:

      (i) $2x - 5y + 1 = 0$, $x + y + 4 = 0$,
      (ii) $7x - 4y + 1 = 0$, $x - y + 1 = 0$,
      (iii) $ax + by - 1 = 0$, $bx + ay - 1 = 0$.

1.4.3   In each of the following cases determine whether P, Q, R are collinear, and if so find the line through them:

(i) $P = (1, -3)$, $Q = (-1, -5)$, $R = (2, -2)$,
(ii) $P = (3, 1)$, $Q = (-1, 2)$, $R = (19, -3)$,
(iii) $P = (4, 3)$, $Q = (-2, 1)$, $R = (1, 2)$.

1.4.4   Find the value of λ for which $P = (3, 1)$, $Q = (5, 2)$, $R = (\lambda, -3)$ are collinear.

1.4.5   Show that for any choice of a, b the points $(a, 2b)$, $(3a, 0)$, $(2a, b)$, $(0, 3b)$ are collinear.

1.4.6   In each of the following cases show that the given lines are concurrent:

(i) $3x - y - 2 = 0$, $5x - 2y - 3 = 0$, $2x + y - 3 = 0$,
(ii) $2x - 5y + 1 = 0$, $x + y + 4 = 0$, $x - 3y = 0$,
(iii) $7x - 4y + 1 = 0$, $x - y + 1 = 0$, $2x - y = 0$.

1.4.7   Find the unique value of λ for which the lines $x - 3y + 3 = 0$, $x + 5y - 7 = 0$, $2x - 2y - \lambda = 0$ are concurrent.

1.5 Parametrized Lines

So far we have viewed lines as sets of points in the plane, defined by a single equation. The next step is to take a different viewpoint, and think of lines as 'parametrized' in a natural way. It is a small step, but it develops into a different viewpoint of the subject.

Lemma 1.3 *Let $P = (p_1, p_2)$, $Q = (q_1, q_2)$ be distinct points on a line $ax + by + c = 0$. For any constant t the point $Z(t) = (x(t), y(t))$, where $x(t)$, $y(t)$ are defined below, also lies on the line*

$$x(t) = (1 - t)p_1 + tq_1, \qquad y(t) = (1 - t)p_2 + tq_2. \qquad (1.4)$$

Conversely, any point $Z = (x, y)$ on the line has this form for some constant t.

Proof The first claim follows from the following identity, as both expressions in braces are zero

$$ax(t) + by(t) + c = (1 - t)\{ap_1 + bp_2 + c\} + t\{aq_1 + bq_2 + c\}.$$

Conversely, for any point $Z = (x, y)$ on L the relation (1.2) is satisfied. Thus the rows of the matrix are linearly dependent, and there are constants s, t for

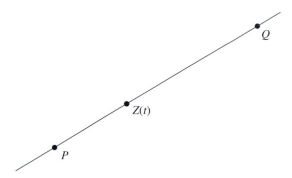

Fig. 1.2. Parametrization of a line

which the following relation holds. The relations (1.4) follow immediately

$$(x, y, 1) = s(p_1, p_2, 1) + t(q_1, q_2, 1).$$

$\square$

The relations (1.4) are the *standard parametrization* of the line with respect to the points P, Q. The mental picture is that the line is traced by a moving particle having the position $Z(t)$ at time t: at time $t = 0$ the particle is at $P = Z(0)$, and at time $t = 1$ it is at $Q = Z(1)$.

Example 1.6 Consider the line $2x - 3y + 3 = 0$. By inspection the line passes through the points $P = (0, 1)$, $Q = (3, 3)$ giving the parametrization $x(t) = 3t$, $y(t) = 2t + 1$. A different choice gives rise to a different parametrization. For instance $P = (-3, -1)$, $Q = (6, 5)$ produces $x(t) = 3(3t - 1)$, $y(t) = 6t - 1$.

The proof of Lemma 1.3 shows that the zero set of any line is infinite, since different values of t correspond to different points $Z(t)$ on the line. The *midpoint* of the line is the point R with parameter $t = 1/2$, i.e. the point

$$R = \frac{P + Q}{2}.$$

Exercises

1.5.1 In each of the following cases find the standard parametrization of the line L relative to the points P, Q:

(i) $L = x - 2y - 5$, $P = (3, -1)$, $Q = (7, 1)$,
(ii) $L = 3x + y - 1$, $P = (3, -8)$, $Q = (-1, -2)$.

1.5.2   In each of the following cases find parametrizations (with integral coefficients if possible) for the given lines:

     (i) $x + 3y - 7 = 0$,   (iv) $2x + 6y - 5 = 0$,
     (ii) $3x - 4y - 13 = 0$,   (v) $2x - 3y + 1 = 0$,
     (iii) $7x - 3y - 8 = 0$,   (vi) $5x - 3y + 1 = 0$.

1.5.3   Find equations for the following parametrized lines:

     (i) $x = 2 + 3t$, $y = -1 + 4t$,
     (ii) $x = \frac{1}{2} + \frac{3}{4}t$, $y = -3 + t$,
     (iii) $x = -3 - t$, $y = 1 - 2t$.

1.5.4   Show that the parametrized lines $x = 2 + 3t$, $y = -1 + 4t$ and $x = -4 + 6t$, $y = -9 + 8t$ coincide.

1.5.5   Find the three intersections of the following parametrized lines:

     (i) $x = 2 + 3t$, $y = 1 - t$,
     (ii) $x = 4 + 4t$, $y = 1 - 2t$,
     (iii) $x = -3 - t$, $y = 2 + 3t$.

1.5.6   Show that any non-vertical line has a parametrization of the form $x(t) = t$, $y(t) = \alpha + \beta t$, and that any non-horizontal line has a parametrization of the form $x(t) = \gamma + \delta t$, $y(t) = t$.

1.6 Pencils of Lines

By the *pencil of lines* spanned by two distinct lines L, M we mean the set of all lines of the form $\lambda L + \mu M$, where λ, μ are constants, not both zero. The key *intersection property* of a pencil is that any two distinct lines L', M' in it have the same intersection as L, M. To this end, write

$$L' = \lambda L + \mu M, \qquad M' = \lambda' L + \mu' M.$$

Since L', M' are distinct, the vectors (λ, μ), (λ', μ') are linearly independent, and by linear algebra the relations $L' = 0$, $M' = 0$ are equivalent to $L = 0$, $M = 0$. That establishes the claim.

    The first geometric possibility for the pencil of lines spanned by L, M is that L, M intersect at a point P. Then, by the intersection property any line in the pencil passes through P, and we refer to the pencil of lines *through P*. Any line $ax + by + c = 0$ through $P = (p, q)$ must satisfy $ap + bq + c = 0$, so can be written in the form

$$a(x - p) + b(y - q) = 0. \tag{1.5}$$

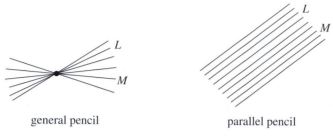

general pencil   parallel pencil

Fig. 1.3. Pencils of lines

Example 1.7 Let L, M be distinct lines in the pencil of all lines through a point P. We claim that *any* line N through P is in the pencil, so has the form $N = \lambda L + \mu M$ for some constants λ, μ. As above, we can write

$$\begin{cases} L(x, y) = a(x - p) + b(y - q) \\ M(x, y) = c(x - p) + d(y - q) \\ N(x, y) = e(x - p) + f(y - q). \end{cases}$$

Since L, M have different directions, the vectors (a, b), (c, d) are linearly independent, and by linear algebra form a basis for the plane. Thus there exist unique constants λ, μ (not both zero) for which the displayed relation below holds. It follows that $N = \lambda L + \mu M$, as was required

$$(e, f) = \lambda(a, b) + \mu(c, d).$$

Example 1.8 In the pencil of lines through $P = (p, q)$ there is a unique vertical line $L(x, y) = x - p$, and a unique horizontal line $M(x, y) = y - q$. By the previous example, any line in the pencil is a linear combination of L, M, as is illustrated by equation (1.5).

The second geometric possibility for the pencil of lines spanned by L, M is that L, M are parallel, so by the intersection property *any* two distinct lines in the pencil are parallel. We call this a *parallel pencil* of lines, and think of it as a limiting case of a general pencil, where all the lines 'pass through' the same point at infinity. In such a pencil all the lines have the *same* direction $-b : a$, so it makes sense to refer to the parallel pencil in that direction.

Example 1.9 Any line in the direction $-b : a$ has an equation of the form $N = 0$, where $N = ax + by + c$ for some constant c. Conversely any line of this form must be in the pencil. Suppose indeed that $L = ax + by + l$,

$M = ax + by + m$ are two distinct lines in the pencil. Then we can write $N = \lambda L + \mu M$, where

$$\lambda = \frac{c - m}{l - m}, \qquad \mu = \frac{l - c}{l - m}.$$

Finally, it is worth noting one small difference between the two geometric possibilities described above. Consider the pencil of lines spanned by two distinct lines L, M. In the case when the lines intersect at a point P, *every* expression $\lambda L + \mu M$ is automatically a linear function, so defines a line. However, when the lines are parallel, there is a unique ratio $\lambda : \mu$ for which $\lambda L + \mu M$ fails to be a linear function. For instance, in the parallel pencil spanned by $L(x, y) = x$, $M(x, y) = 2x - 1$, the expression $2L - M = 1$ fails to be linear.

Exercise

1.6.1   Show that the pencil of lines spanned by the lines $2x + 3y - 8$, $4x - 7y + 10$ coincides with the pencil spanned by $3x + 4y - 11$, $2x - 5y + 8$.

2

The Euclidean Plane

The material of the previous chapter lay wholly within the context of linear algebra. In this chapter we introduce the Euclidean structure on the plane, the central concept on which the rest of this text depends. It is that structure which enables us to introduce length, angle, and distance. The key technical fact is the Cauchy inequality, leading directly to the Triangle inequality, and the introduction of angle. In the final section we introduce the distance between a point and a line, the key to the focal constructions of conics in Chapter 4.

2.1 The Scalar Product

The plane is endowed with its standard *Euclidean structure*. By this we mean that for any two vectors $Z_1 = (x_1, y_1)$, $Z_2 = (x_2, y_2)$ we have the standard *scalar product* (or *dot product*) defined by the relation

$$Z_1 \bullet Z_2 = x_1 x_2 + y_1 y_2.$$

The basic properties (Exercise 2.1.1) of the scalar product are listed below:

S1: $Z \bullet Z \geq 0$ with equality if and only if $Z = 0$.
S2: $Z \bullet W = W \bullet Z$.
S3: $Z \bullet (\lambda W) = \lambda(Z \bullet W)$.
S4: $Z \bullet (W + W') = Z \bullet W + Z \bullet W'$.

S2 is referred to as the *symmetry* property. Properties S3, S4 together say that $\bullet$ is linear in its second argument: by symmetry, it is also linear in its first argument, and for that reason $\bullet$ is said to be *bilinear*. Two vectors Z, W are *perpendicular* when $Z \bullet W = 0$.

Example 2.1 Let Z, W be vectors with $W \neq 0$. We claim that there is a unique scalar λ with the property that the vectors $Z' = Z - \lambda W$, W are perpendicular.

Indeed, our requirement is that

$$0 = Z' \bullet W = (Z - \lambda W) \bullet W = Z \bullet W - \lambda (W \bullet W).$$

That gives the unique solution

$$\lambda = \frac{Z \bullet W}{W \bullet W}.$$

We call the vector λW the *component of Z parallel to W*, and the vector $Z' = Z - \lambda W$ the *component of Z perpendicular to W*.

Exercise

2.1.1   Starting from the definition of the scalar product, establish the properties S1, S2, S3, S4.

2.2 Length and Distance

Property S1 of the scalar product is expressed by saying that the scalar product is *positive definite*. In view of this property it makes sense to define the *length* of a vector $Z = (x, y)$ to be

$$|Z| = \sqrt{x^2 + y^2}.$$

Throughout this book we will use the following fundamental properties L1, L2, L3 of the length function. The property L1 is an immediate consequence of S1 above: however L2 and L3 require proof, representing the next step in our development

L1: $|Z| = 0$ if and only if $Z = 0$
L2: $|Z \bullet W| \le |Z||W|$ (The Cauchy Inequality)
L3: $|Z + W| \le |Z| + |W|$ (The Triangle Inequality)

Lemma 2.1 *For any two vectors Z, W in the plane we have the relation* $|Z \bullet W| \le |Z||W|$. *(The Cauchy Inequality.)*

Proof When $Z = 0$ the LHS is zero, and the inequality is satisfied. We can therefore assume that $Z \ne 0$, so $Z \bullet Z > 0$. Set $\lambda = Z \bullet W / Z \bullet Z$. Then λZ represents the component of W parallel to Z and $W - \lambda Z$ is the component

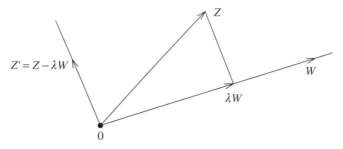

Fig. 2.1. Components of a vector

perpendicular to Z. (Figure 2.1.) Then

$$0 \le |W - \lambda Z|^2 = (W - \lambda Z) \bullet (W - \lambda Z)$$
$$= W \bullet W - 2\lambda(Z \bullet W) + \lambda^2(Z \bullet Z)$$
$$= W \bullet W - \lambda(Z \bullet W)$$
$$= |W|^2 - \frac{(Z \bullet W)^2}{|Z|^2}.$$

The result follows on multiplying through by $|Z|^2$ and taking positive square roots. □

Lemma 2.2 *For any two vectors Z, W in the plane we have the relation* $|Z + W| \le |Z| + W|$. *(The Triangle Inequality.)*

Proof The Cauchy Inequality yields the following series of relations, from which the result follows on taking positive square roots

$$|Z + W|^2 = (Z + W) \bullet (Z + W) = Z \bullet Z + 2(Z \bullet W) + W \bullet W$$
$$\le |Z|^2 + 2|Z \bullet W| + |W|^2 \le |Z|^2 + 2|Z||W| + |W|^2$$
$$= (|Z| + |W|)^2.$$

□

We define the *distance* between two points U, V in the plane to be the scalar $|U - V|$. The following basic properties of the distance function are immediate from L1, L2, L3

M1: $|U - V| = 0$ if and only if $U = V$
M2: $|U - V| = |V - U|$
M3: $|U - V| \le |U - W| + |W - V|$.

Note that distance is *invariant under translation*, in the sense that for any vector W the distance between U, V equals that between their translates $U + W$, $V + W$.

Exercises

2.2.1   A vector Z is *unit* when $|Z| = 1$. Let U, V be unit vectors. Show that $U + V, U - V$ are perpendicular.

2.2.2   Establish the following *parallelogram law* for any vectors U, V

$$|U + V|^2 + |U - V|^2 = 2|U|^2 + 2|V|^2.$$

2.2.3   The length of a vector was expressed in terms of the scalar product. Conversely, show that the scalar product can be expressed in terms of the length via the *polarization identity*

$$U \bullet V = \frac{1}{2}\{|U|^2 + |V|^2 - |U - V|^2\}.$$

2.2.4   Let A, B be distinct points on a line L, and let e be a positive constant. By parametrizing L show that when $e \neq 1$ there exist two distinct points P for which $PA = ePB$, and that when $e = 1$ there is just one.

2.3 The Concept of Angle

Given two non-zero vectors V, V' we can write the Cauchy Inequality in the following form

$$-1 \le \frac{V \bullet V'}{|V||V'|} \le 1.$$

Looking at the graph of the cosine function[1] we see that there is therefore a unique constant α with $0 \le \alpha \le \pi$ for which

$$\cos \alpha = \frac{V \bullet V'}{|V||V'|}. \tag{2.1}$$

We call α the *angle* between the vectors V, V'. Thus when V, V' are unit vectors the scalar product is just the cosine of the angle between them. It is usual to refer to angles in the range $0 \le \alpha < \pi/2$ as *acute*, to the angle $\alpha = \pi/2$ as a *right angle*, and to angles in the range $\pi/2 < \alpha \le \pi$ as *obtuse*. Thus two vectors V, V' are perpendicular when the angle between them is a right angle.

---

[1] More formally, we are using the fact from elementary analysis that given any constant t with $-1 \le t \le 1$ there is a unique constant α in the interval $0 \le \alpha \le \pi$ for which $\cos \alpha = t$.

Example 2.2 Let A, B, C be non-zero vectors with $C = A - B$, and let α be the angle between A, B. Expanding the expression for $|C|^2$ we obtain the *cosine rule* of school trigonometry

$$|C|^2 = |A|^2 - 2|A||B|\cos\alpha + |B|^2.$$

A special case arises when A, B are perpendicular, so the angle α is a right angle: the cosine rule then reduces to the familiar *Pythagoras Theorem*

$$|C|^2 = |A|^2 + |B|^2.$$

Our next step is to extend the idea of angle from vectors to lines. First a preliminary observation. We can write any line $L = ax + by + c$ in the form $L = V \bullet Z + c$, where $V = (a, b)$ and $Z = (x, y)$. Thus when $c = 0$ (or equivalently, the line passes through the origin) the zero set of the line is the set of vectors Z perpendicular to V. The vector V can always be chosen to be a *unit* vector, by dividing L through by the length of V: in that case the line is said to be in *canonical form*. There are therefore two canonical forms for a line, each obtained from the other by multiplying through by -1. For instance the line $L = 3x + 4y - 2$ has the canonical forms

$$\frac{3x + 4y - 2}{5}, \qquad \frac{-3x - 4y + 2}{5}.$$

Consider now two lines L, L' defined by the linear functions $ax + by + c$, $a'x + b'y + c'$. The angle between the vectors $V = (a, b)$, $V' = (a', b')$ is by definition the unique constant α with $0 \le \alpha \le \pi$ for which $\cos\alpha = t$, where

$$t = \frac{V \bullet V'}{|V||V'|}.$$

It is important to realise that *this expression depends on the choice of linear functions defining the lines*, since multiplication of the linear functions by non-zero constants multiplies t by ± 1.

Lemma 2.3 *Let α, β be the unique angles in the range $0 \le \alpha, \beta \le \pi$ for which $\cos\alpha = t$, $\cos\beta = -t$. Then $\alpha + \beta = \pi$.*

Proof Using the Difference Formula for the cosine function we have $\cos\alpha = -\cos\beta = \cos(\pi - \beta)$, and since the angles α, $\pi - \beta$ both lie in the interval $0 \le \theta \le \pi$ they are equal.[2]   □

The angles α, β of this lemma are the *angles* between the lines L, L'. (Figure 2.2.) Generally, one of the angles between two lines is acute, and the other obtuse: the only exception is when both angles are right angles.

---

[2] We are using another fact from elementary analysis, namely that in the interval $0 \le \theta \le \pi$ the cosine function is strictly decreasing.

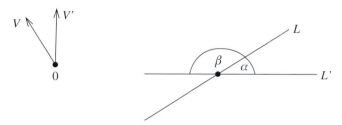

Fig. 2.2. Angles between two lines

Example 2.3 We will determine the angles between the two lines $L = 4x - y - 2$, $L' = 3x - 5y + 1$. In this example $V = (4, -1)$, $V' = (3, -5)$ and the angles θ between the lines are determined by

$$\pm \cos \theta = \frac{V \bullet V'}{|V||V'|} = \frac{17}{\sqrt{17}\sqrt{34}} = \frac{1}{\sqrt{2}}.$$

It now follows from elementary trigonometry that the required angles are $\alpha = \pi/4$, $\beta = 3\pi/4$.

An exceptional case arises when both the angles between L, L' are right angles, in which case we say that the lines are *perpendicular*. The condition for $L = ax + by + c$, $L' = a'x + b'y + c'$ to be perpendicular is that the vectors $V = (a, b)$, $V' = (a', b')$ should be perpendicular

$$aa' + bb' = 0. \tag{2.2}$$

A *direction vector* (or just a *direction*) for a line $L = ax + by + c$ is any non-zero vector perpendicular to $V = (a, b)$. In particular, the vector $V^{\perp} = (-b, a)$ is a direction, and any other direction is a scalar multiple of that vector.

Example 2.4 The line joining the two distinct points $P = (p_1, p_2)$, $Q = (q_1, q_2)$ is $ax + by + c = 0$, where

$$a = p_2 - q_2, \qquad b = -(p_1 - q_1), \qquad c = p_1 q_2 - p_2 q_1.$$

Thus a direction vector associated to the line is $P - Q$. For instance, the line joining $P = (2, -1)$, $Q = (2, -2)$ is $3x + 4y - 2 = 0$, having direction vector $P - Q = (-4, 3)$.

Example 2.5 Let P, Q be distinct points. A point Z is *equidistant* from P, Q when the distances from Z to P, Q are equal. That is equivalent to

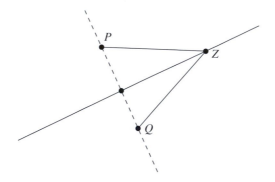

Fig. 2.3. The perpendicular bisector

$$|P - Z|^2 = |Q - Z|^2.$$

The set of points equidistant from P, Q is called the *perpendicular bisector* of the line segment joining P, Q. (Figure 2.3.) We claim that the perpendicular bisector is a line. Expanding both sides of the above relation we obtain

$$2(Q - P) \bullet Z = |Q|^2 - |P|^2.$$

That is the equation of a line perpendicular to the direction vector $Q - P$ of the line joining P, Q. Note that the midpoint of the line segment joining P, Q lies on the perpendicular bisector.

Exercise

2.3.1   In each of the following cases determine the angles between the given lines L, L':

(i) $L = 2x - y - 1,$ $L' = x - 2y + 1,$
(ii) $L = 3x + 4y - 7,$ $L' = 2x - 3y - 8,$
(iii) $L = 2x - y + 3,$ $L' = 3x - y + 2.$

2.4 Distance from a Point to a Line

The next result will be of significance in Chapter 8, when we discuss focal constructions of conics. It is exceptional in this text in that it uses calculus methods to find the stationary points of a function.

Lemma 2.4 *Let $L = ax + by + c$ be a line, and $R = (x, y)$ a point. There is a unique point Q on L for which the distance between Q, R is a minimum d,*

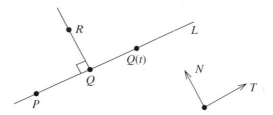

Fig. 2.4. Projection of a point on to a line

given by the formula

$$d^2 = \frac{L(x, y)^2}{a^2 + b^2}.$$ (2.3)

Proof The vector $N = (a, b)$ is perpendicular to L, and $T = (-b, a)$ is a direction vector for L. Choose any point P on L. Then L is defined by the formula $L(Z) = N \bullet (Z - P)$, and parametrized as $Q(t) = P + tT$. The square of the distance from R to $Q(t)$ is given by the function

$$f(t) = (R - Q(t)) \bullet (R - Q(t)).$$

This function has a stationary point when its derivative with respect to the variable t vanishes. Differentiating, we obtain

$$f'(t) = -2Q'(t) \bullet (R - Q(t)).$$

This expression vanishes if and only if $Q'(t) = T$ is perpendicular to $R - Q(t)$, i.e. when the following relation holds, determining a unique value of t, hence a unique point Q

$$0 = T \bullet (R - P - tT) = T \bullet (R - P) - t(T \bullet T).$$

Moreover, it is a strict minimum of the function, since the second derivative $f''(t) = 2(T \bullet T)$ is positive. For that value $Q = P + tT = R + sN$ for some constant s, so $R - P = -sN + tT$. Taking the scalar product of both sides with N, we obtain

$$-s = \frac{N \bullet (R - P)}{N \bullet N} = \frac{L(x, y)}{a^2 + b^2}.$$

Finally, we obtain the required formula (2.3) by observing that the minimum value d is given by

$$d^2 = (R - Q) \bullet (R - Q) = -sN \bullet -sN = s^2(a^2 + b^2).$$

$\square$

The proof shows that there is a unique point Q on L for which the line joining Q, R is perpendicular to L. That point Q is the *projection* of R on to L. Equivalently, Q can be described as the *nearest point* to R on L. The *distance RL* from R to L is defined to be $|R - Q|$, where Q is the projection of R on to L.

Example 2.6 We will determine the projection of $R = (2, 3)$ on to the line $L = 4x + 3y - 7$, and the distance between them. In this example $T = (-3, 4)$, and we could choose $P = (1, 1)$, so L is parametrized as $Q(t) = (1 - 3t, 1 + 4t)$. The condition for T to be perpendicular to $R - Q(t)$ is that

$$0 = (-3, 4) \bullet (1 + 3t, 1 + 4t) = 5(1 - 5t).$$

That gives $t = 1/5$ so the projection of R on to L is $Q = (2/5, 9/5)$. The formula (2.3) then gives $RL = 2$, since

$$RL^2 = \frac{L(2, 3)^2}{4^2 + 3^2} = \frac{(4.2 + 3.3 - 7)^2}{4^2 + 3^2} = \frac{10^2}{25} = 4.$$

Example 2.7 Let $L = ax + by + c$, $L' = ax + by + c'$ be parallel lines, and let $R = (\alpha, \beta)$ be a point on L', so $a\alpha + b\beta + c' = 0$. Then (2.3) tells us that the distance d from R to L is determined by

$$d^2 = \frac{L(\alpha, \beta)^2}{a^2 + b^2} = \frac{(a\alpha + b\beta + c)^2}{a^2 + b^2} = \frac{(c - c')^2}{a^2 + b^2}.$$

This expression depends only on L, L': it does not depend on the choice of point R. For that reason d is defined to be the *distance* between the parallel lines L, L'. Note that in the special case when L, L' are in canonical form the distance is just $d = |c - c'|$.

Exercises

2.4.1   Show that two non-vertical lines $y = px + q$, $y = p'x + q'$ are perpendicular if and only if $pp' = -1$.

2.4.2   Find the equation of the line perpendicular to $x + 2y - 4$ through the intersection of the lines $3x + 4y - 8$, $2x - 5y + 3$.

2.4.3   Find the line through the points $P = (1, 1)$, $Q = (4, 3)$, and the perpendicular line through $R = (-1, 1)$.

2.4.4   In each of the following cases find the projection of the point R on to the line L, and hence determine the distance from R to L:

(i) $L = 3x + 4y - 5,$ $R = (2, 6),$
(ii) $L = 5x + 12y - 20,$ $R = (2, 3),$
(iii) $L = 3x + 4y - 5,$ $R = (4, 2).$

2.4.5   Let L, L' be parallel lines. Show that there is a unique line L'' parallel to L, L' and equidistant from them. The line L'' is the *parallel bisector* of L, L'.

3

Circles

Apart from lines there are few geometric objects of such fundamental importance to geometry as circles. In school mathematics a 'circle' is viewed as the set of points (x, y) which are a constant positive distance r from a fixed point (α, β), so defined by an equation

$$C(x, y) = (x - \alpha)^2 + (y - \beta)^2 - r^2 = 0. \qquad (3.1)$$

We will pursue the same line of thought set out in Chapter 1 for lines. Thus we think in terms of the formula $C(x, y)$ rather than the set of points it defines. That makes it more natural to present a wider concept of 'circle' than is familiar from school mathematics, better suited to a systematic development. The resulting 'circles' represent a useful stepping stone to general conics since (despite their simplicity) circles illustrate some of their vagaries. For instance, their zero sets can be infinite, a single point or empty. In the first case we show that the zero set determines the 'circle' up to constant multiples, using the basic idea expounded in Section 1.3. In the final section we look at the way in which circles intersect lines. That can viewed as an introduction to a recurring theme of this text, namely the way in which conics intersect lines, a topic we will expand upon in Chapter 4.

3.1 Circles as Conics

The expression in (3.1) is an example of a *quadratic function* on the plane, a function Q given by a formula of the following form, where the *coefficients* a, $2h, b, 2g, 2f, c$ are such that at least one of a, b, h is non-zero

$$Q(x, y) = ax^2 + 2hxy + by^2 + 2gx + 2fy + c. \qquad (\star)$$

Two quadratic functions Q, Q' are *scalar multiples* of each other when there exists a real number $\lambda \neq 0$ with $Q' = \lambda Q$: the resulting equivalence classes are known as *conics*. For instance, the three quadratic functions below all define the same conic

$$x^2 + y^2 - 1, \qquad 2x^2 + 2y^2 - 2, \qquad -x^2 - y^2 + 1.$$

The conic arising from a quadratic function Q will be denoted by the same symbol: on the occasions when we do need to draw a distinction we will deliberately use the term 'quadratic function'. Associated to any quadratic function Q is its *zero set*

$$\{(x, y) \in \mathbb{R}^2 : \ Q(x, y) = 0\}.$$

Note that any scalar multiple of Q has the same zero set, so the concept makes perfect sense for conics. Instead of saying that $P = (x, y)$ is a point in the zero set, we shall (for linguistic variety) say that P *lies on* Q, or that Q *passes through P*.

3.2 General Circles

A *circle* is defined to be a conic $(\star)$ with the property that $a = b$ and $h = 0$, so can be written in the following form with $a \neq 0$

$$a(x^2 + y^2) + 2gx + 2fy + c. \tag{3.2}$$

Dividing through by the common coefficient of x^2, y^2 we see that any circle is defined by a quadratic function C in the *canonical form* displayed below

$$C(x, y) = x^2 + y^2 - 2\alpha x - 2\beta y + \gamma. \tag{3.3}$$

The *centre* of C is the point (α, β). For a fixed centre the nature of the zero set depends on the value of the constant γ. To make this explicit, write C in the following form, where $k = \gamma - \alpha^2 - \beta^2$

$$C(x, y) = (x - \alpha)^2 + (y - \beta)^2 + k. \tag{3.4}$$

We call C a *real circle* when $k < 0$, a *point circle* when $k = 0$, and a *virtual circle* when $k > 0$.

Example 3.1 The case of greatest physical interest is that of the real circle. In that case we can write $k = -r^2$, where r is a positive constant, the *radius* of C. Note that the centre of a real circle C does not lie on the circle. Also, the zero set of C is infinite, since the *standard parametrization* below shows that it contains infinitely many points

$$x(\theta) = \alpha + r \cos\theta, \qquad y(\theta) = \beta + r \sin\theta. \tag{3.5}$$

Example 3.2 The situation changes dramatically for the point and virtual circles. In the case $k = 0$ of a point circle we have $C(x, y) \geq 0$ for all x, y with equality if and only if $x = \alpha$, $y = \beta$: thus the zero set of a point circle is a single point, namely the centre. And in the case $k > 0$ of a virtual circle we have $C(x, y) > 0$ for all x, y, so the zero set of a virtual circle is empty.

Virtual circles reveal a disturbing feature of conics. For distinct positive values k_1, k_2 of k in (3.4) we obtain virtual circles C_1, C_2 whose zero sets coincide (both are empty) though C_1, C_2 are not scalar multiples of each other: thus *a conic is not necessarily determined by its zero set*. That is one reason why we have to proceed more carefully with conics than we did with lines.

Exercises

3.2.1   In each of the following cases determine the centre and type of the given circle, and in the case of a real circle its radius:

     (i) $x^2 + y^2 - 6x + 2y - 39$,
     (ii) $4x^2 + 4y^2 - 4x - 5y + 1$,
     (iii) $x^2 + y^2 - 2ax$.

3.2.2   Let $A = (1, 0)$, $B = (-1, 0)$ and let $\lambda > 0$. Show that for $\lambda \neq 1$ the set of points $P = (x, y)$ for which $|PA| = \lambda |PB|$ is the zero set of a real circle centred on the x-axis. Investigate how the circle changes as λ varies.

3.3 Uniqueness of Equations

We observed above that virtual circles are not determined by their zero sets. However, real circles do have this property. To prove this we adopt the same simple-minded approach as we did for lines. The circle C displayed in (3.3) has three arbitrary coefficients. We would reasonably expect these to be determined by three conditions, for instance that C should pass through three general points. Indeed, that proves to be the case.

Theorem 3.1 *Through any three non-collinear points* $P_1 = (x_1, y_1)$, $P_2 = (x_2, y_2)$, $P_3 = (x_3, y_3)$ *there is a unique real circle*

$$C(x, y) = x^2 + y^2 - 2\alpha x - 2\beta y + \gamma.$$

Proof The condition for the circle to pass through the three points is that the following relations hold, representing a system of three linear equations in the three unknowns α, β, γ

$$\begin{cases} 2\alpha x_1 + 2\beta y_1 - \gamma = x_1^2 + y_1^2 \\ 2\alpha x_2 + 2\beta y_2 - \gamma = x_2^2 + y_2^2 \\ 2\alpha x_3 + 2\beta y_3 - \gamma = x_3^2 + y_3^2. \end{cases}$$

By linear algebra these equations have a unique solution provided the 3×3 matrix of coefficients has a non-zero determinant. That however is the condition of Example 1.4 for the points P_1, P_2, P_3 to be non-collinear. $\quad\square$

Thus *a real circle is determined by its zero set*, meaning that if two real circles C, C' have the same zero sets they are scalar multiples of each other: we have only to pick three distinct points in the common zero set, and apply the result. That justifies the time-honoured practice of referring to the *equation* $C = 0$ of a circle C. Strictly, that is an abbreviation for the zero set of C, but since the zero set determines C (up to scalar multiples) it is not too misleading. Of course, the question arises as to what extent general conics are determined by their zero sets. The answer is as follows.

Theorem 3.2 *Let Q, Q' be conics having the same zero set. Then Q, Q' coincide, provided the common zero set is infinite.* (The Uniqueness Theorem.)

The proof is a natural extension of that for Theorem 3.1. For the reader's convenience it is delayed till Chapter 17. The Uniqueness Theorem belongs to a central theme in algebraic geometry, developed in *EGAC*: its practical import is that *provided the zero set is infinite* it makes sense to refer to the 'conic' $Q(x, y) = 0$. Strictly speaking, that is just a shorthand notation for the zero set of a quadratic polynomial Q, but the Uniqueness Theorem guarantees that any other quadratic polynomial Q' with the same zero set is a scalar multiple of Q, so defines the same conic.

Exercises

3.3.1   In each of the following cases find the circle through the given points P, Q, R:

(i) $P = (0, 0)$, $Q = (a, 0)$, $R = (0, b)$,
(ii) $P = (1, 0)$, $Q = (2, 3)$, $R = (-1, -1)$,
(iii) $P = (-1, 1)$, $Q = (-1, 3)$, $R = (2, 4)$.

3.3.2   A set of points is *concyclic* when there exists a circle passing through every point in the set. Show that the points $(-3, 11)$, $(5, 9)$, $(8, 0)$, $(6, 8)$ are concyclic, and that they lie on a circle with centre $(-1, 2)$.

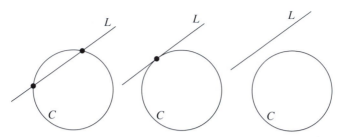

Fig. 3.1. How circles intersect lines

3.4 Intersections with Lines

In this section we introduce a recurring theme of this text, the question of how conics intersect lines. For the moment we confine ourselves to the more specific question of how a circle C intersects lines. A sketch suggests that C will intersect a line L in at most two points. That is easily verified. Suppose L is parametrized as $x(t) = u + tX$, $y(t) = v + tY$. Substituting in (3.3) we obtain a quadratic equation in t

$$0 = C(u + tX, v + tY) = pt^2 + qt + r.$$

In any example the coefficients are easily calculated. For the moment, all that is important is that $p = X^2 + Y^2$, so is non-zero. Thus the quadratic has two distinct roots, one repeated root, or no roots. It follows that C meets L in two distinct points, just one point, or not at all. A *chord* of C is a line L meeting C in two points, called the *ends* of the chord: exceptionally the intersection is a single point, in which case we say that L *touches* C at the point with parameter t. A *diameter* of a circle is a chord passing through the centre.

Example 3.3 A sketch suggests that any line through the centre of a real circle is a diameter. Let us verify that for the real circle (3.1) with centre (α, β) and radius r. Let L be a line through the centre, parametrized in the following form for some fixed angle θ

$$x(t) = \alpha + t\cos\theta, \qquad y(t) = \beta + t\sin\theta.$$

Substituting in (3.1) gives $t^2 = r^2$ with distinct solutions $t = \pm r$. Substituting in the displayed relations we obtain the points of the standard parametrization (3.5) corresponding to the angles $\theta, \theta + \pi$.

Example 3.4 Consider the intersections of the circle $x^2 + y^2 = 25$ and the line $x - 7y + 25 = 0$. Parametrizing the line as $x(t) = 3 + 7t$, $y(t) = 4 + t$

and substituting in the circle we get a quadratic $50t(t + 1) = 0$. The roots are $t = 0, -1$ corresponding to the two intersections $(3, 4)$, $(-4, 3)$. If instead we take the line $3x + 4y - 25 = 0$ parametrized as $x(t) = 3 - 4t$, $y(t) = 4 + 3t$ the quadratic is $25t^2 = 0$ with a repeated root $t = 0$, so the line touches the circle at the sole intersection $(3, 4)$.

Exercise

3.4.1   Let $P = (a, b)$, $P' = (a', b')$ be distinct points. Show that there is a unique circle for which P, P' are the ends of its diameter, and having the equation displayed below. For fixed P, describe the locus of points P' for which the circle passes through the origin

$$(x - a)(x - a') + (y - b)(y - b') = 0.$$

3.5 Pencils of Circles

By the *pencil of circles* spanned by two distinct circles C, D we mean the set of all circles of the form $\lambda C + \mu D$ where λ, μ are constants, not both zero. The concept is wholly analogous to that of a pencil of lines, introduced in Section 1.6, and has the same key *intersection property* that any two distinct elements C', D' in the pencil have the same intersection as C, D: the proof is identical to the line case. Note that there is a unique exceptional ratio $\lambda : \mu$ for which $\lambda C + \mu D$ *fails* to be a quadratic function: for all other ratios it is a quadratic function defining a circle. We can be more explicit by writing C, D in their canonical forms

$$\begin{cases} C(x, y) = x^2 + y^2 - 2\alpha x - 2\beta y + \gamma \\ D(x, y) = x^2 + y^2 - 2\alpha' x - 2\beta' y + \gamma'. \end{cases} \tag{3.6}$$

With those choices the exceptional ratio is $-1 : 1$, and the corresponding element of the pencil is

$$L(x, y) = 2(\alpha - \alpha')x + 2(\beta - \beta')y - (\gamma - \gamma'). \tag{3.7}$$

Provided C, D have distinct centres, L is a line, known as the *radical axis* of the circles. Exercise 3.5.8 shows that any two distinct circles in the pencil $\lambda C + \mu D$ have the same radical axis L. By the intersection property, the intersections of C, D coincide with those of C or D with the line L. The results of the previous section show that a circle intersects a line in two distinct points, just one point, or not at all. We can therefore conclude that any two distinct circles intersect in two distinct points, just one point, or not at all. In particular, the circles

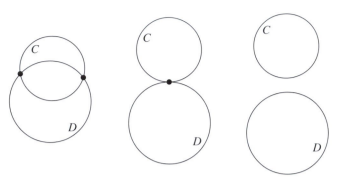

Fig. 3.2. Three ways in which circles can intersect

intersect in a single point exactly when L touches C, D at the same point: in that case we say that C, D *touch* at that point. A special case arises when C, D are *concentric*, i.e. have the same centre. In that case $\alpha = \alpha'$, $\beta = \beta'$, and $L(x, y) = \gamma' - \gamma$ is a non-zero constant function: it follows that distinct concentric circles do not intersect. However obvious that fact may be visually, it does require formal proof!

Example 3.5 The radical axis of the circles displayed below is the line $x = 0$. Substituting $x = 0$ in the first circle gives $y^2 + 1 = 0$, having no solutions. The radical axis does not therefore meet the first circle, and the circles do not intersect

$$x^2 + y^2 - 3x + 1 = 0, \qquad 2x^2 + 2y^2 - 7x + 2 = 0.$$

Example 3.6 Recall that the perpendicular bisector L of the line segment joining two distinct points $P = (\alpha, \beta)$, $Q = (\alpha', \beta')$ was introduced in Example 2.5 as the locus of points equidistant from P, Q. It is the line with equation

$$2(Q - P) \bullet z = |Q|^2 - |P|^2.$$

We can think of L in another way. Let C, D be the point circles with centres P, Q so having the canonical forms (3.6) with $\gamma = \alpha^2 + \beta^2$, $\gamma' = \alpha'^2 + \beta'^2$. Then comparison of the displayed equation with (3.7) shows that L is the radical axis of C, D.

Example 3.7 Let C, D be non-concentric circles. We claim that the centres of the circles $\lambda C + \mu D$ all lie on the *centre line*, the line M joining the centres of C, D. To prove this, take C, D to be in their canonical forms (3.6), and define constants s, t with $s + t = 1$ by

$$s = \frac{\lambda}{\lambda + \mu}, \qquad t = \frac{\mu}{\lambda + \mu}.$$

With these choices the canonical form for $\lambda C + \mu D$ is given by the expression

$$x^2 + y^2 - 2(s\alpha + t\alpha')x - 2(s\beta + t\beta')y + (sc + tc').$$

It follows that its centre is the point $s(\alpha, \beta) + t(\alpha', \beta')$ on the parametrized line joining the centres (α, β), (α', β') of C, D. The formula (1.1) shows that the centre line is

$$M(x, y) = (\beta - \beta')x - (\alpha - \alpha')y + (\alpha\beta' - \alpha'\beta) = 0.$$

The condition (2.2) for perpendicularity shows that for any two non-concentric circles C, D the radical axis L is perpendicular to the centre line M. The next example illustrates this general fact.

Example 3.8 Consider the special case of the preceding example when the radical axis is the y-axis. Looking at (3.7) we see that is the case if and only if $\beta = \beta' = 0$ and $c = c'$. In that case the centre line is the x-axis, and the canonical form of the circle $\lambda C + \mu D$ has the following shape for a constant ν depending on λ, μ

$$x^2 + y^2 - 2\nu x + c = 0.$$

The nature of the family depends on the sign of c, and is best understood by writing the equation in the form

$$(x - \nu)^2 + y^2 = \nu^2 - c.$$

When c is negative (say $c = -k^2$) all the circles are real, with radius $\geq k$ the value $\nu = 0$ giving the circle of minimal radius k, centre the origin; moreover, all the circles have common points $(0, \pm k)$ on the radical axis. When c is positive (say $c = k^2$) the circles are real only in the range $-k < \nu < k$ and fail to cut the radical axis: outside this range they are virtual, except for the two point circles at $(\pm k, 0)$. The remaining case $c = 0$ is transitional, with all the circles real, touching the radical axis at the origin. The common points of the case $c < 0$ coincide at the origin, as do the point circles of the case $c > 0$.

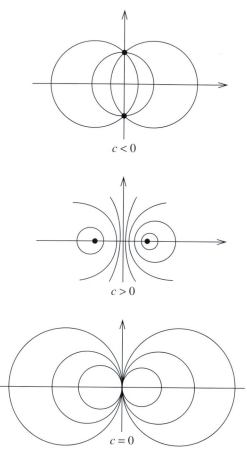

$$c < 0$$

$$c > 0$$

$$c = 0$$

Fig. 3.3. The family of circles in Example 3.8

Exercises

3.5.1   In each of the following cases show that the given line L touches the given circle C, and state the point of contact:

(i) $L = 2x + 3y,$ $C = x^2 + y^2 + 3y + 2x,$
(ii) $L = 3x - 4y - 10,$ $C = x^2 + y^2 + 2x - 6y,$
(iii) $L = 5x - 12y - 45,$ $C = x^2 + y^2 + 16x - 14y - 56.$

3.5.2   In each of the following cases find the radical axis of the given circles, and their intersections:

(i) $x^2 + y^2 - x - 3y + 3,$ $x^2 + y^2 + 8x - 6y - 3,$
(ii) $x^2 + y^2 - 4x + 6y + 8,$ $x^2 + y^2 - 10x - 6y + 14,$
(iii) $x^2 + y^2 - 6x - 6y - 14,$ $x^2 + y^2 - 2.$

3.5.3   Find the circles touching the lines $x = 0$, $y = 0$, $x = 2a$, where a is a positive constant.

3.5.4   Find the circles touching the lines $x = 2$, $y = 5$, $3x - 4y = 10$.

3.5.5   Find the circles which touch both the coordinate axes and pass through the point $(6, 3)$.

3.5.6   Show that the three radical axes associated to three circles with distinct centres are concurrent or parallel.

3.5.7   Show that if two of the radical axes associated to three circles with distinct centres coincide, then all three coincide.

3.5.8   Let C, D be circles with distinct centres and radical axis the line L. Show that any two distinct circles C', D' in the pencil $\lambda C + \mu D$ also have distinct centres, and the same radical axis.

3.5.9   Show that for $\lambda \neq -1$ the formula below defines a circle, that the centre lies on the line $3x + y - 5 = 0$, and that the radical axis of any two circles in the family is the line $x + y = 0$. Further, show that the circles are real if and only if $-3 < \lambda < 0$, and find the two point circles in the family

$$(x - 1)^2 + (y - 2)^2 + \lambda(x^2 + y^2 + 2x + 5) = 0.$$

4

General Conics

In this chapter we take a first look at more general conics than circles, before launching ourselves into more detailed considerations. One of our long-term objectives will be to separate out general conics into a small number of types, distinguished by their underlying geometry. We start by introducing the reader to the 'standard' conics which will play a dominant role in this text. They are not simply examples of conics: they turn out to be *models* of the physically most important conics, in a sense made precise in Chapter 15. The qualitative form of their zero sets can be determined by looking carefully at the way in which they intersect the pencils of horizontal and vertical lines, and offers insight into the computer generated illustrations. Like lines and circles the 'standard' conics admit natural parametrizations, of practical value in elucidating their geometry.

We will need simple and effective means for distinguishing one type of conic from another. As a first step in this direction we introduce three easily calculated 'invariants' of a general conic, namely the trace invariant τ, the delta invariant δ, and the discriminant Δ. All three are easily calculated expressions in the coefficients, from which we can read off useful geometric information. However, their true significance does not appear till the final chapter, where it is shown that they are 'invariant' in a strictly defined sense. It is the delta invariant and the discriminant which yield the most useful information, and provide us with a first crude subdivision of conics into types. By contrast, the trace invariant will play only a minor role.

Section 4.4 considers the broad question of how a general conic Q intersects a line L. The situation is analogous to that of the previous chapter, where we saw that circles intersect lines in at most two points. However, when dealing with a general conic Q there is one important exception, namely that every point on L may be a point of Q: the function of the final section is to establish the Component Lemma, that L is then actually a factor of Q.

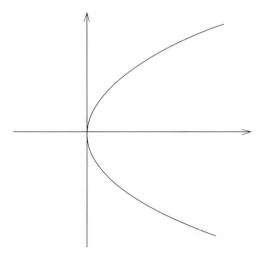

Fig. 4.1. A standard parabola

4.1 Standard Conics

The circles of Chapter 3 were our first examples of conics. In the following examples we introduce three families of 'standard' conics which play a major role in the subject. All the standard conics have infinite zero sets, so by the Uniqueness Theorem are determined by their zero sets. An important feature of standard conics is that they have at least one 'axis' of symmetry. In Chapter 7 we will formally introduce the concept of 'axis' for a general conic, and explain how to find the axes. It will turn out that any conic has at least one 'axis', and that the axes of the standard conics are precisely those described in the following examples.

Example 4.1 Let a be a positive constant. The *standard* parabola with *modulus* a is the conic with equation

$$y^2 = 4ax. \tag{4.1}$$

Figure 4.1 illustrates the zero set, traced using a computer program. However, we can predict its qualitative form by looking at its intersections with the pencils of horizontal and vertical lines. A vertical line $x = c$ meets Q exactly twice for $c > 0$, just once for $c = 0$, and not at all for $c < 0$. By contrast, any horizontal line $y = d$ meets Q at exactly one point. The line $y = 0$ is the *axis* of the standard parabola: the curve has an evident symmetry in that line.

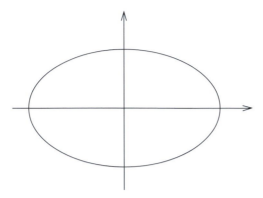

Fig. 4.2. A standard ellipse

Example 4.2 Let a, b be constants with $0 < b < a$. The *standard* real ellipse with *moduli* a, b is the conic with the equation

$$\frac{x^2}{a^2} + \frac{y^2}{b^2} = 1. \tag{4.2}$$

Figure 4.2 illustrates the zero set. A vertical line $x = c$ meets Q exactly twice for $-a < c < a$, just once for $c = a, -a$, and not at all for $c < -a$ or $c > a$. Likewise, horizontal lines $y = d$ meet Q exactly twice for $-b < d < b$, just once for $d = b, -b$, and not at all for $d < -b$ or $d > b$. The coordinate axes are the *axes* of the standard ellipse: the curve has evident symmetries in each line. The *major* axis $y = 0$ meets the ellipse at the points $(\pm a, 0)$ distant $2a$ apart, whilst the *minor* axis $x = 0$ meets the ellipse at the points with $(0, \pm b)$ distant $2b$ apart.

The standard ellipses do not include real circles since the moduli are subject to the constraint that $0 < b < a$, so the coefficients of x^2, y^2 cannot be equal. That is a deliberately imposed constraint, for a reason that will become clear in Chapter 8. Nevertheless, it is profitable to think of a real circle (of radius a, and centre the origin) as the limiting case of standard ellipses as $b \to a$.

Example 4.3 Let a, b be positive constants. The *standard* hyperbola with *moduli* a, b is the conic with the equation

$$\frac{x^2}{a^2} - \frac{y^2}{b^2} = 1. \tag{4.3}$$

Figure 4.3 illustrates the zero set. A vertical line $x = c$ meets Q exactly twice for $c > a$ or $c < -a$, just once for $c = a, -a$, and not at all for $-a < c < a$. However, any horizontal line $y = d$ meets Q exactly twice. Thus the zero set

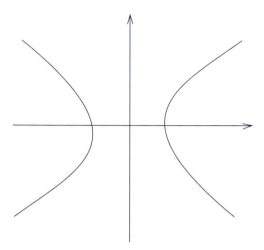

Fig. 4.3. A standard hyperbola

of a standard hyperbola splits into two 'branches', namely the *positive branch*, defined by $x \geq a$, and the *negative branch*, defined by $x \leq -a$. The coordinate axes are the *axes* of the standard hyperbola: the curve has evident symmetries in each line. The *transverse* axis $y = 0$ meets the hyperbola at two points with $x = \pm a$, distant $2a$ apart, whilst the *conjugate* axis $x = 0$ fails to meet the hyperbola.

4.2 Parametrizing Conics

We have already met parametrizations of lines and real circles. They are useful devices, often easier to deal with than the defining function. For that reason it is helpful to extend the concept to general conics. The function of this section is to present a fairly minimal account, sufficient for the rest of this text. A *parametrization* of a conic Q comprises two smooth functions $x(t)$, $y(t)$ defined on an open interval I satisfying the following relation for all t in I

$$Q(x(t), y(t)) = 0.$$

Thus I is a set of real numbers t (the *parameters*) satisfying an inequality $a < t < b$, where we allow $a = -\infty$, $b = \infty$. The meaning of the term 'smooth' is that at every parameter t both $x(t)$ and $y(t)$ have derivatives of all orders. In all our examples it will be self evident that $x(t)$, $y(t)$ are 'smooth' in this sense. Mostly, the domain I will be the whole real line $\mathbb{R}$: only infrequently, when the domain is not $\mathbb{R}$, will we specify it. The *image* of the parametrization is the set of all points $(x(t), y(t))$ with t in I, so is contained in the zero set of Q. There is no reason to suppose that every point in the zero set will be in the image: for instance, both $x(t)$ and $y(t)$ might be constant, with image a single point. The parametrization is *regular* when for any parameter t at least one of the derivatives $x'(t)$, $y'(t)$ is non-zero: a parameter t for which $x'(t) = 0$, $y'(t) = 0$ is said to be *irregular*. Here are some explicit parametrizations of the standard conics; in later chapters we will describe techniques for parametrizing general conics.

Example 4.4 Consider the standard parabola $y^2 = 4ax$ with modulus $a > 0$. Each line in the pencil of lines $y = 2at$ parallel to the x-axis meets the parabola just once, at the point where $x = at^2$. In this way we obtain the regular parametrization

$$x(t) = at^2, \qquad y(t) = 2at. \tag{4.4}$$

Example 4.5 Consider the standard ellipse with moduli a, b for which $0 < b < a$. The x-coordinate of any point on the ellipse satisfies the inequality $-a \le x \le a$, so can be written $x = a \cos t$ for some t. Substituting in the equation of the ellipse we obtain $y = \pm b \sin t$. The '$+$' option gives a regular parametrization of the ellipse, in terms of the *eccentric angle* t, tracing the ellipse anticlockwise

$$x(t) = a \cos t, \qquad y(t) = b \sin t. \tag{4.5}$$

The '$-$' option gives another regular parametrization of the ellipse, tracing the curve clockwise. Of course the argument applies equally well when $b = a$: in that case we recover the parametrization of the circle of radius a, centre the origin in Example 3.1.

The parametrization of a standard hyperbola is more thought provoking than that of a standard ellipse. One approach is to replace the trigonometric functions by hyperbolic functions.

Example 4.6 Let (x, y) be a point satisfying the equation of the standard hyperbola with moduli a, b. Glancing at the graph of the sinh function we see that we can write $y = b \sinh t$ for a unique real number t. Then, substituting in the equation we see that $x = \pm a \cosh t$. The positive and negative branches are then parametrized as

$$x(t) = \pm a \cosh t, \qquad y(t) = b \sinh t. \tag{4.6}$$

The choice of the hyperbolic functions in this example is by no means mandatory. Indeed, the only property of the sinh function we have used is that it is smooth and bijective. We could just as well take the tangent function, which has the same property.

Example 4.7 For any point (x, y) on the standard hyperbola with moduli a, b we can write $y = b \tan t$ for some t with $-\pi < 2t < \pi$. Substituting for y in the equation we obtain $x = \pm a \sec t$, leading to two further parametrizations of the positive and negative branches

$$x(t) = \pm a \sec t, \qquad y(t) = b \tan t.$$

Exercises

4.2.1   Show that the standard parametrization (3.5) of a real circle is regular. What is its image?

4.2.2   Show that $x(t) = 2r \cos^2 t$, $y(t) = 2r \sin t \cos t$ is a regular parametrization of the real circle of radius r, centre $(r, 0)$.

4.2.3   For the parametrization $x(t) = at^2$, $y(t) = 2at$ of the standard parabola, show that the chord through the points with parameters t, t' has equation

$$(t + t')(y - 2at) = 2(x - at^2).$$

4.2.4   Show that $x(t) = \cos^4 t$, $y(t) = \sin^4 t$ is a parametrization of the parabola $(x - y - 1)^2 = 4y$. Find the irregular parameters, and their images on the parabola.

4.2.5   Show that the lines joining points with the same parameter t on the ellipse $x(t) = a \cos t$, $y(t) = b \sin t$ and the hyperbola $x(t) = a \sec t$, $y(t) = b \tan t$ comprise a pencil.

4.3 Matrices and Invariants

One feature of a quadratic function is that it can be succinctly described by a matrix. We adopt the standard notation introduced in the last chapter

$$Q(x, y) = ax^2 + 2hxy + by^2 + 2gx + 2fy + c. \qquad (\star)$$

To the general quadratic function $(\star)$ we associate the 3×3 symmetric *matrix*

Table 4.1. *Non-degenerate classes*

name	δ	Δ
ellipses	$\delta > 0$	$\Delta \neq 0$
parabolas	$\delta = 0$	$\Delta \neq 0$
hyperbolas	$\delta < 0$	$\Delta \neq 0$

displayed below

$$A = \begin{pmatrix} a & h & g \\ h & b & f \\ g & f & c \end{pmatrix}. \tag{4.7}$$

Writing z for the row vector $z = (x, y, 1)$, and z^T for its transposed column vector, we see that Q can be written usefully in terms of its associated matrix

$$Q(x, y) = zAz^T. \tag{4.8}$$

The matrix A gives rise to three expressions which play a significant role in the study of conics, namely the *invariants* τ, δ, Δ defined as follows

$$\tau = a + b, \qquad \delta = \begin{vmatrix} a & h \\ h & b \end{vmatrix}, \qquad \Delta = \begin{vmatrix} a & h & g \\ h & b & f \\ g & f & c \end{vmatrix}. \tag{4.9}$$

Bear in mind that the invariants are associated to the quadratic function Q, as opposed to the conic. When we multiply Q by a scalar $\lambda \neq 0$ we multiply all its coefficients by λ, and hence τ, δ, Δ by λ, λ^2, λ^3 respectively. Although that changes the invariants, it does not change the equalities and inequalities

$$\tau = 0, \quad \delta = 0, \quad \Delta = 0, \quad \tau \neq 0, \quad \delta > 0, \quad \delta < 0, \quad \Delta \neq 0.$$

The *discriminant* Δ is the dominant invariant, with the *delta* invariant δ playing a lesser role, and the *trace* invariant τ sitting quietly in the wings. The conics of greatest interest are *non-degenerate*, meaning that the discriminant is non-zero: of these there are three broad types, displayed in Table 4.1. Ellipses split naturally into two classes, namely *real* ellipses having non-empty zero sets, and *virtual* ellipses having empty zero sets; only the former are geometrically significant.

Example 4.8 We leave the reader to check that the invariants of the general circle C in canonical form (3.2) are given by the relations

$$\tau = 2, \qquad \delta = 1, \qquad \Delta = c - \alpha^2 - \beta^2.$$

It follows that any real circle is in the real ellipse class, any virtual circle is in the virtual ellipse class, and any point circle is degenerate. Likewise, the reader will readily check that the standard parabolas, ellipses, and hyperbolas are indeed parabolas, real ellipses, and hyperbolas in the sense of the definition just given.

Exercises

4.3.1 Show that there does not exist a non-zero quadratic polynomial Q for which the invariants $\tau = 0$, $\delta = 0$.

4.3.2 In each of the following cases, calculate the invariants of the given conic, and hence determine its class:

(i) $5x^2 + 6xy + 5y^2 - 4x + 4y - 4$,

(ii) $4x^2 - 4xy + y^2 - 10y - 19 = 0$,

(iii) $2x^2 - xy - 3y^2 + 4x - 1 = 0$.

4.3.3 In each of the following cases calculate the invariants of the given conic in terms of λ, find the values for which it degenerates, and the conic class when it is non-degenerate:

(i) $2y(x - 1) + 2\lambda(x - y)$,

(ii) $\lambda(x^2 + y^2) - (x - 1)^2$,

(iii) $x^2 + t(t + 1)y^2 + 2 - 2\lambda xy + 2x$.

4.4 Intersections with Lines

We gained some feel for the geometry of the *standard* conics by looking carefully at their intersections with lines. That suggests it will be profitable to understand better how *general* conics intersect lines. It is a fruitful interaction, leading to useful insights. Indeed it turns out to be a fundamental idea in studying general plane curves, developed in *EGAC* and *EGCD*. We have already observed in Section 3.4 that a circle meets a line L in at most two points. Our next objective is to show that a general conic Q has the same property, save for the exceptional case when *every* point on L is a point of Q.

Lemma 4.1 *Let L be a line, and Q a conic. Then L intersects Q in two distinct points, one point, or no points; otherwise, every point on L is a point on Q.*

Proof Let L be a line in the direction (X, Y) through a point (u, v), parametrized as $x(t) = u + tX$, $y(t) = v + tY$, and write

$$Q(x, y) = ax^2 + 2hxy + by^2 + 2gx + 2fy + c. \qquad (\star)$$

The values of t for which $(x(t), y(t))$ lies on Q are given by $\phi(t) = 0$, where

$$\phi(t) = Q(x(t), y(t)) = Q(u + tX, v + tY). \qquad (4.10)$$

Since $\phi(t)$ is obtained by substituting linear terms in t into a quadratic function, it will be a quadratic in t, so have the form

$$\phi(t) = pt^2 + qt + r. \qquad (4.11)$$

Provided at least one coefficient in (4.11) does not vanish, there are either two distinct roots, one root, or no roots: otherwise, all three coefficients vanish, and *every* value of t is a root. $\qquad \square$

We call $\phi(t)$ the *intersection quadratic*. The reader is invited to verify the explicit formulas for the coefficients displayed below: they show that p depends solely on (X, Y), that q depends on both (X, Y) and (u, v), and that r depends solely on (u, v)

$$\begin{cases} p = a^2 X^2 + 2h XY + b^2 Y^2 \\ q = Q_x(u, v)X + Q_y(u, v)Y \\ r = Q(u, v). \end{cases} \qquad (4.12)$$

Example 4.9 We will determine the intersections of the conic Q and the line L defined below

$$Q(x, y) = x^2 + 2xy - 3y^2 + 8y + 3, \qquad L(x, y) = 2x - y + 3.$$

The line L passes through the point $(-1, 1)$, and has direction $(1, 2)$, so can be parametrized as $x(t) = -1 + t$, $y(t) = 1 + 2t$. Substituting in Q, we find that $\phi(t) = -7(t^2 - 1)$. Thus $\phi(t)$ has zeros $t = 1, -1$ and Q intersects L at the points $(0, 3)$, $(-2, -3)$ with these parameters.

It is possible for the intersection quadratic to have a repeated root. To pursue this possibility a little further let us recall basic facts about quadratic equations. By the Factor Theorem of school algebra, t_0 is a root of the quadratic $\phi(t) = 0$ if and only if $(t - t_0)$ is a factor of $\phi(t)$. Recall that t_0 is a *repeated* root when $(t - t_0)^2$ is a factor of $\phi(t)$, in which case we say that L *touches* Q at the point with parameter t_0. In particular, when $\phi(t)$ is identically zero (all its coefficients are zero) *every* value t_0 is a repeated root of $\phi(t) = 0$.

Example 4.10 Consider the intersections of the conic Q and the line L defined below

$$Q(x, y) = x^2 + y^2 - 5, \qquad L(x, y) = 2x + y - 5.$$

The line L passes throught the point $(3, -1)$, and has direction $(1, -2)$, so can be parametrized as $x(t) = 3 - t$, $y(t) = 2t - 1$. Substituting in Q we get $\phi(t) = 5t^2 - 10t + 5 = 5(t - 1)^2$, with a repeated root $t = 1$. Thus L touches Q at the point on L having parameter $t = 1$, namely the point $(2, 1)$.

Example 4.11 Let L be a line in the direction (X, Y), and let Q be a conic. Then L touches Q when the resulting intersection quadratic $pt^2 + qt + r$ has a repeated zero. Although the coefficients depend on the direction, whether L touches Q is independent of the choice. We can see this as follows. Any other direction for L is a constant multiple $(\lambda X, \lambda Y)$ of the given one, with λ non-zero. And looking at the formulas (4.12) we see that the corresponding intersection quadratic is $p\lambda^2 t^2 + q\lambda t + r$. It remains only to observe that in each case the condition for the quadratic to have a repeated zero is that $q^2 - pr = 0$.

Exercises

4.4.1   In each of the following cases find the intersections of the given line L with the given conic C:

(i) $L = x - 7y + 25$, $Q = x^2 + y^2 - 25$,
(ii) $L = 4x + 3y - 11$, $Q = 2x^2 + 3y^2 - 11$,
(iii) $L = 3x - 2y + 1$, $Q = 6x^2 + 11xy - 10y^2 - 4x + 9y$.

4.4.2   Find the chord of the parabola $y^2 = 8x$ whose midpoint is the point $(2, -3)$.

4.4.3   By considering its intersections with the line $y = 0$, show that the conic Q_t displayed below has a non-empty zero set if and only if $t > 0$

$$Q_t(x, y) = t(x^2 + y^2) - (x - 1)^2.$$

4.5 The Component Lemma

We conclude this chapter by completing our account of how general conics intersect lines. Lemma 4.1 allowed an exceptional situation, namely that a conic Q might intersect a line L at every one of its points. That can certainly happen: for instance the conic $Q(x, y) = xy$ meets the lines $x = 0$, $y = 0$ at every one of their points. A conic Q is *reducible* when there exist lines L, L' for which $Q = LL'$: in that case L, L' are the *components* of Q, and Q is the *joint equation* of L, L'. Otherwise Q is *irreducible*. Each component of a reducible conic meets Q at every one of its points. Note that the zero set of a reducible conic Q is infinite, since it is the union of the zero sets of its components.

Reducible conics Q can be distinguished geometrically by the way in which the components L, L' intersect. According to Example 1.5 there are three possibilities, illustrated by Figure 1.1. The first possibility is that L, L' have distinct directions: in that case we refer to Q as a *real line-pair*, and to the unique point of intersection of L, L' as the *vertex*. Otherwise L, L' have the same direction: when they are parallel Q is a *real parallel line-pair*, and when they coincide Q is a *repeated line*.

Example 4.12 The conic $Q = x^2 - xy - 2y^2 + 2x + 5y - 3$ reduces, since $Q = LL'$, where $L = x + y - 1$, $L' = x - 2y + 3$. Indeed Q is a line-pair since the components have different directions.

The object of this section is to prove the Component Lemma, that the only way in which a line L can meet a conic Q at every point of L is when L is a component. The proof depends on the following technical lemma, that we can 'divide' Q by L to obtain a 'quotient' L' and a 'remainder' J.

Lemma 4.2 *Let Q be a conic, and let L be a line not parallel to the y-axis. There is a unique line L', and a unique quadratic $J(x)$ with*

$$Q(x, y) = L(x, y)L'(x, y) + J(x). \qquad (4.13)$$

Proof Write $L(x, y) = \alpha x + \beta y + \gamma$, $L'(x, y) = \alpha' x + \beta' y + \gamma'$: the hypothesis, that L is not parallel to the y-axis, means that $\beta \neq 0$. We can take Q to be given by $(\star)$. Equating coefficients of the monomials y^2, xy, y in both sides of (4.13) gives three linear equations in α', β', γ'

$$b = \beta\beta', \qquad 2h = \alpha\beta' + \alpha'\beta, \qquad 2f = \beta\gamma' + \beta'\gamma.$$

The determinant of the 3×3 matrix of coefficients is $-\beta^3$, hence non-zero. By linear algebra there is a unique solution α', β', γ', yielding a unique line L'. It remains to note that $J(x)$ is uniquely determined by the requirement that

$$Q(x, 0) = L(x, 0)L'(x, 0) + J(x).$$

$\square$

Assuming instead that L is not parallel to the x-axis, we obtain a similar conclusion, with $J(x)$ replaced by a quadratic $K(y)$. The only difference in the proof is that the line L' is determined by equating the coefficients of x^2, xy, x. Here is the promised pay-off.

Lemma 4.3 *Suppose that every point on the line L lies on a conic Q. Then* $Q = LL'$ *for some line L'. In particular, that is the case when L meets Q in more than two points.* (The Component Lemma.)

Proof We can assume L it is not parallel to the y-axis. By Lemma 4.2 there is a line L' and a quadratic $J(x)$ for which (4.13) holds. For every x there is a unique value of y for which $L(x, y) = 0$, and hence $Q(x, y) = 0$. That means that $J(x) = 0$ for all x: since a non-zero quadratic has ≤ 2 roots, that means J is identically zero, so $Q = LL'$. Finally, if L meets Q in ≥ 3 points, then Lemma 4.1 tells us that every point on L lies on Q, so we reach the same conclusion. $\square$

Example 4.13 The reader is left to check that every point on the line $L(x, y) = 2x - y - 1$ lies on the conic Q below, by substituting $y = 2x - 1$ in $Q(x, y)$ to obtain a zero expression

$$Q(x, y) = 2x^2 + xy - y^2 + x - 2y - 1.$$

By the Component Lemma L is a factor of Q. Indeed $Q = LL'$, where $L'(x, y) = x + y + 1$.

Exercises

4.5.1   Show that the components of a real line-pair Q are perpendicular if and only if the trace invariant τ vanishes.

4.5.2   Let L, L' be lines through the origin having the joint equation $ax^2 + 2hxy + by^2 = 0$. Show that the joint equation of the lines M, M' perpendicular to L, L' is $bx^2 - 2hxy + ay^2 = 0$.

5

Centres of General Conics

A striking feature of a circle is that there is a point (the centre) which does not lie on the circle itself, yet is crucial to understanding the geometry of the curve. The concept of 'centre' is by no means unique to circles. Our first step is to introduce the idea for general conics: that provides the material for Section 5.1. However, general conics do not always have centres, presenting us with one crude way of distinguishing some conics from others. For that remark to be useful we need to have an efficient practical technique to find the centres of a conic, if any. That is the function of Section 5.2. These considerations enable us to distinguish three broad classes of conics, namely those having a unique centre, those having no centre, and those having a line of centres. And that will provide a basis for the classification of conics in Chapter 15.

5.1 The Concept of a Centre

We are used to thinking of the centre of a circle as the point equidistant from the points in its zero set. There is however another approach, capable of generalization. Any line through the centre meets the circle in two distinct points, and the centre is the midpoint of the resulting chord. That suggests how we might extend the concept to general conics. Let $W = (u, v)$ be a fixed point. By *central reflexion in W* we mean the mapping of the plane defined by the rule $(x, y) \to (2u - x, 2v - y)$: a special case arises when W is the origin, and central reflexion is given by $(x, y) \to (-x, -y)$. The key geometric property is that the midpoint of the line joining a point (x, y) to its central reflexion $(2u - x, 2v - y)$ in W is the point W itself. We say that W is a *centre* for a conic Q when the following identity holds

$$Q(x, y) = Q(2u - x, 2v - y). \tag{5.1}$$

Suppose (u, v) is a centre for Q. Then it is an immediate consequence of the definition that a point (x, y) is in the zero set of Q if and only if its central

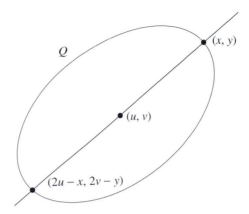

Fig. 5.1. The concept of a centre

reflexion $(2x - u, 2y - v)$ is. However, it is worth remarking that the definition makes sense whether or not the zero set of Q contains points.

Example 5.1 The point (α, β) is a centre of the circle Q below, since it is unchanged when x, y are replaced by $2\alpha - x$, $2\beta - y$. That confirms that the use of the term 'centre' in Chapter 3 is consistent with the above general definition

$$Q(x, y) = (x - \alpha)^2 + (y - \beta)^2 + \gamma.$$

5.2 Finding Centres

A conic Q is *central* when it has at least one centre. We need a practical procedure for deciding whether or not Q is central, and if so finding centres. That is the motivation for the following definition. By *translation* of the plane through a vector (u, v) we mean a mapping of the plane defined by a formula $(X, Y) \rightarrow (x, y)$, where $x = X + u$, $y = Y + v$. On a geometric level, translation represents a sliding of the plane in the direction (u, v). Given a conic Q and a vector (u, v) we define the *translate* of Q through (u, v) to be the conic R obtained from Q via the substitutions $x = X + u$, $y = Y + v$, or symbolically

$$R(X, Y) = Q(X + u, Y + v).$$

It is worth noting two practicalities. The first is that, although the linear and constant terms in Q change under translation, the quadratic terms remain unchanged. And the second is that the constant term in $R(X, Y)$ is obtained by setting $X = 0$, $Y = 0$ so is $Q(u, v)$.

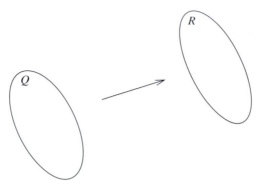

Fig. 5.2. A translate of a conic

Example 5.2 For the conic Q defined below we will determine the translate through $(3, -2)$

$$Q(x, y) = 2x^2 + 3y^2 - 12x + 12y + 24.$$

Substituting $x = X + 3$, $y = Y - 2$ we find after a certain amount of calculation that the translate is

$$R(X, Y) = 2X^2 + 3Y^2 - 6.$$

Lemma 5.1 *A point (u, v) is a centre for a conic $Q(x, y)$ if and only if the origin $(0, 0)$ is a centre for the translated conic*

$$R(X, Y) = Q(X + u, Y + v).$$

Proof The condition for (u, v) to be a centre for Q is that we have the identity (5.1). Replacing x, y by $X + u$, $Y + v$ we obtain the following identity, expressing the fact that $(0, 0)$ is a centre for R

$$R(X, Y) = Q(X + u, Y + v) = Q(-X + u, -Y + v) = R(-X, -Y).$$

$\square$

That reduces our problem to that of finding a practical criterion for the origin to be a centre for the general conic

$$Q(x, y) = ax^2 + 2hxy + by^2 + 2gx + 2fy + c. \qquad (\star)$$

Lemma 5.2 *The origin is a centre for a general conic $(\star)$ if and only if the coefficients of the linear terms x, y are both zero.*

Proof The origin is a centre of Q if and only if the following conics coincide.

Since at least one of a, h, b is non-zero, that happens if and only if $f = 0$, $g = 0$, i.e. the coefficients of the linear terms x, y are zero

$$\begin{cases} Q(x, y) = ax^2 + 2hxy + by^2 + 2gx + 2fy + c \\ Q(-x, -y) = ax^2 + 2hxy + by^2 - 2gx - 2fy + c. \end{cases}$$

$\square$

We can now put the bits together to obtain a practical method for finding all the centres of any given conic.

Lemma 5.3 *The centres (u, v) of a general conic $(\star)$ are the solutions of the linear equations*

$$au + hv + g = 0, \qquad hu + bv + f = 0. \tag{5.2}$$

Proof According to Lemma 5.1 the centres of a conic $Q(x, y)$ are those points (u, v) for which the origin is a centre of $Q(x + u, y + v)$. And by Lemma 5.2 these are the points (u, v) for which the coefficients of the linear terms in $Q(x + u, y + v)$ are zero. The coefficients are found by replacing x, y by $x + u$, $y + v$ in $Q(x, y)$ and then collecting all terms in x, and all terms in y. We leave the reader to verify that the coefficients of x, y are respectively $2(au + hv + g)$, $2(hu + bv + f)$. $\square$

There is another way of stating this result which is perhaps easier to remember in practice. Write Q_x, Q_y for the partial derivatives of Q with respect to the variables x, y. The reader will verify that

$$Q_x(x, y) = 2(ax + hy + g), \qquad Q_y(x, y) = 2(hx + by + f).$$

Thus the centres of a conic Q are the solutions of the linear equations

$$Q_x(x, y) = 0, \qquad Q_y(x, y) = 0. \tag{5.3}$$

Example 5.3 For the standard parabola $Q(x, y) = y^2 - 4ax$ with $a > 0$ centres are the solutions of $Q_x = -4a = 0$, $Q_y = 2y = 0$. These equations are inconsistent, so standard parabolas have no centre.

The analogy with the circle is by no means perfect. The next example shows how the general concept of a 'centre' can differ from the familiar one for a circle.

Example 5.4 Consider conics of the form $Q = ax^2 + by^2 + c$. Centres are given by $Q_x = 2ax = 0$, $Q_y = 2by = 0$. Provided a, b are both non-zero the origin is the only centre, lying in the zero set if and only if $c = 0$. In

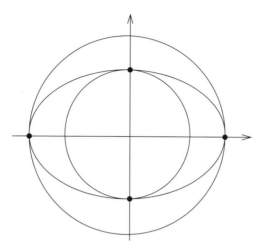

Fig. 5.3. Auxiliary circles of ellipse

particular, the origin is the unique centre of any standard ellipse or hyperbola. When $a = 0$ there is a line of centres $y = 0$; and likewise when $b = 0$ is zero there is a line of centres $x = 0$. Moreover, all these cases are independent of whether the zero set of Q is empty.

Two conics Q, Q' are *concentric* when they have a common unique centre. In the case of two circles that agrees with the definition given in Chapter 3. Here is a situation giving rise to concentric conics which are not necessarily circles.

Example 5.5 The two vertices of a standard ellipse Q on a given axis are necessarily equidistant from the centre. The two circles concentric with a standard ellipse and passing through two of the vertices are the *auxiliary circles* associated to Q: the *minor* auxiliary circle is that of smaller radius, and the *major* auxiliary circle is that of larger radius. Likewise, the two vertices on the transverse axis of a standard hyperbola are equidistant from the centre, and the circle through them concentric with the hyperbola is its *auxiliary circle*.

Exercises

5.2.1   Determine the translate of the conic Q below under the vector $W = (-1, 1)$

$$Q = 17x^2 - 12xy + 8y^2 + 46x - 28y + 17.$$

5.2.2   In each of the following cases decide whether the given conic has a centre, and if so find all the centres:

(i) $25x^2 + 9y^2 - 72y - 81 = 0$,
(ii) $5x^2 + 6xy + 5y^2 - 4x + 4y - 4 = 0$,
(iii) $x^2 - 4xy + 4y^2 + 10x - 8y + 13 = 0$.

5.2.3   In each of the following cases decide whether the given conic has a centre, and if so find all the centres:

(i) $x^2 + xy - 2y^2 + x - y = 0$,
(ii) $x^2 - 4xy + 3y^2 + 2x = 0$,
(iii) $x^2 - 2xy + y^2 + 2x - 4y + 3 = 0$.

5.2.4   In each of the following cases show that given conic is central for all t, and that the centres are collinear:

(i) $2y(x - 1) + 2t(x - y)$,
(ii) $t(x^2 + y^2) - (x - 1)^2$,
(iii) $x^2 + t(t + 1)y^2 + 2 - 2txy + 2x$. $(t \neq 0)$.

5.2.5   Let P be the intersection of two distinct lines L, M. Show that P is the unique centre of the real line-pair $Q = LM$.

5.2.6   Let N be the parallel bisector of two distinct parallel lines L, M. Show that N is a line of centres for the real parallel line-pair $Q = LM$.

5.3 Geometry of Centres

Although Lemma 5.3 is presented as a practical technique for finding centres, it also yields essential geometric information, summarized by the following result.

Theorem 5.4 *The general conic Q defined by $(\star)$ has a unique centre, a line of centres, or no centre. There is a unique centre if and only if $\delta \neq 0$. And if there is a line of centres then $\Delta = 0$.*

Proof By Lemma 5.3 the centres of Q are the solutions of the linear system of equations

$$ax + hy + g = 0, \qquad hx + by + f = 0. \qquad (5.4)$$

By linear algebra, there is a unique solution, a line of solutions, or no solution: moreover, there is a unique solution if and only if the determinant of the 2×2 matrix of coefficients is non-zero, i.e. $\delta = ab - h^2 \neq 0$. If there is a line of centres, the vectors (a, h, g), (h, b, f) are linearly dependent. Thus the

first two rows of the matrix (4.1) are linearly dependent, and its determinant
$\Delta = 0$. $\square$

Example 5.6 We have already observed that standard parabolas have no centres. In fact that is true of general parabolas. By definition, a parabola is a conic
Q for which $\delta = 0$ and $\Delta \neq 0$. The fact that $\delta = 0$ means that Q has either a
line of centres, or no centre. However, in the former case Theorem 5.4 ensures
that $\Delta = 0$, a contradiction. Thus Q has no centres.

In a given example the equations (5.2) can be solved *ad hoc*. However, there
is value in having general formulas. To this end we introduce a useful notation.
Write $A, B, C, \ldots$ for the cofactors of $a, b, c, \ldots$ in the matrix (4.1) defining
the discriminant Δ. Thus

$$\begin{cases} A = bc - f^2, & B = ca - g^2, & C = ab - h^2 \\ F = gh - af, & G = hf - bg, & H = fg - ch. \end{cases} \tag{5.5}$$

Suppose now that there is a unique centre. Then solving the equations (5.2)
explicitly, for instance via Cramer's Rule, we find that the coordinates of the
centre are

$$u = \frac{G}{C}, \qquad v = \frac{F}{C}. \tag{5.6}$$

Here is a useful payoff. Suppose we have a conic Q with a unique centre.
When we translate the centre to the origin, to obtain a translated conic R, the
quadratic terms do not change, and the linear terms will be absent, so it is
only the constant term that requires calculation. The next result determines the
constant term explicitly in terms of the invariants for Q. That means that R can
be calculated without knowing the coordinates of the centre.

Lemma 5.5 *Assume the general conic Q has a unique centre (u, v). Then the
constant term in the translated conic below is Δ/δ*

$$R(X, Y) = Q(X + u, Y + v).$$

Proof The constant term in $R(X, Y)$ is obtained by setting $X = 0$, $Y = 0$
giving $Q(u, v)$. Now Q can be written

$$Q(x, y) = x(ax + hy + g) + y(hx + by + f) + (gx + fy + c). \tag{5.7}$$

Since u, v satisfy the equations (5.2) we obtain the following equalities, the
second using the formulas for the coordinates of the centre, and the third on

expanding the determinant of the matrix (4.1) by its last row or column

$$Q(u, v) = gu + fv + c = \frac{gG + fF + cC}{C} = \frac{\Delta}{\delta}.$$

□

Example 5.7 The conic Q defined below has a non-zero delta invariant, hence a unique centre. We will not write out the coordinates of the centre, since they are not relevant to the example

$$Q(x, y) = x^2 - 2xy + 5y^2 + 2x - 10y + 1.$$

The matrix of Q has two rows which are scalar multiples of each other, so $\Delta = 0$. It follows immediately that the conic obtained by translating the centre to the origin is

$$R(X, Y) = X^2 - 2XY + 5Y^2.$$

Exercise

5.3.1　In each of the following cases show that the conic Q has a unique centre. Without finding the centre, determine the conic obtained by translating the centre to the origin:

(i)　$Q = 2x^2 - 3xy - 2y^2 + 2x + 11y - 13$,
(ii)　$Q = 5x^2 + 6xy + 5y^2 - 4x + 4y - 4$,
(iii)　$Q = x^2 - 4xy + 3y^2 + 2x$.

5.4 Singular Points

An exceptional situation which may arise for a conic Q is that one of its points is a centre. That gives rise to the following definition. A *singular point* of a conic Q is a centre in its zero set: and Q is *singular* when it has at least one singular point. By (5.3) the condition for Q to be singular is that there exists a point (x, y) satisfying the simultaneous equations

$$Q(x, y) = 0, \qquad Q_x(x, y) = 0, \qquad Q_y(x, y) = 0. \qquad (5.8)$$

Note first that any conic whose zero set comprises a single point W is singular, since according to the definitions W is automatically a centre. Here is a rather different example.

Example 5.8 The conic $Q(x, y) = x^2 - 4y^2 - 2x + 8y - 3$ is singular. The reader is left to check that Q has a unique centre, namely the point $(1, 1)$, and that the centre lies on Q. The geometry underlying this example is very simple. The conic Q is reducible, since we can write $Q = LM$ with $L = x - 2y + 1$, $M = x + 2y - 3$, and the centre is the unique intersection of L, M.

The next result shows that this example is typical, in the following precise sense.

Lemma 5.6 *Let Q be a singular conic whose zero set has more than one point. Then Q is reducible, and its singular points are the intersections of its components.*

Proof Since Q is singular there is a centre W lying in its zero set. And since the zero set of Q comprises more than one point, there is at least one point Z distinct from W in the zero set. As W is a centre, the central reflexion Z' of Z in W is also in the zero set of Q. It follows that the line L through Z, Z', W meets Q in three distinct points, so by the Component Lemma is a component of Q. Let L' be the other component, and write $L = ax + by + c$, $L' = a'x + b'y + c'$. By (5.8) singular points of Q have to satisfy the simultaneous equations

$$LL' = 0, \qquad a'L + aL' = 0, \qquad b'L + bL' = 0.$$

The first relation tells us that $L = 0$ or $L' = 0$. And in either case the other two relations show that both L, L' must vanish, so the singular points are the intersections of the components L, L'. □

To decide whether a conic Q is singular it appears that we must first find its centres, and then check whether at least one of them lies on Q. However, in most cases we can avoid this lengthy computation, using the following fact.

Lemma 5.7 *Suppose the conic Q has a unique centre. Then the centre is singular on Q if and only if $\Delta = 0$.*

Proof Let (u, v) be the centre. It is singular on Q if and only if it lies on Q. And that is the case if and only if $(0, 0)$ lies on the translated conic

$$R(X, Y) = Q(X + u, Y + v).$$

However $(0, 0)$ lies on R if and only if the constant term in R is zero, which by Lemma 5.5 is equivalent to $\Delta = 0$. □

Example 5.9 The conic $Q(x, y) = x^2 - 2xy + 5y^2 + 2x - 10y + 1$ has delta invariant $\delta = 4$, and is non-singular since $\Delta = -36$. We could have achieved the same result by checking that $(0, 1)$ is the only centre, and observing that it does not lie on Q: however, it is easier to calculate the invariants.

Although singular points are exceptional features of a conic, they represent the beginning of a profitable train of geometric thought. The natural generalizations of conics are the algebraic curves studied at length in EGAC. And for such curves the concept of a 'singular point' turns out to be of fundamental importance in understanding their geometry.

6

Degenerate Conics

The conics of greatest geometric significance are the non-degenerate ones, having a non-zero discriminant. Although the geometry of degenerate conics is uninteresting, they are not without significance as transitional types in *families* of conics. That is not the only justification for devoting space to degenerate conics. There are useful facts and techniques associated to them which are worth setting out properly. Section 6.1 introduces a particularly simple class of degenerate conics called binary quadratics, just one small step away from the familiar quadratics of school mathematics, and spells out the mechanics of handling them. That sets the scene for a more careful study of reducible conics in Section 6.2. The key facts are that reducible conics are automatically degenerate, and come close to being characterized by the sign of their delta invariant: moreover, there is an entirely practical procedure for finding the components of a reducible conic. By way of illustration we introduce pencils of conics, and ask how to find degenerate conics within the pencil, a question directly relevant to the focal constructions of Chapter 8. And in the final section we continue the theme by discussing the perpendicular bisectors of a real line-pair, the simplest illustration of a central idea of the subject, namely the axes of a conic.

6.1 Binary Quadratics

In this section we look at a class of degenerate conics, including both reducible and irreducible types. A *binary quadratic* is a conic of the following form

$$Q(x, y) = ax^2 + 2hxy + by^2. \tag{6.1}$$

Binary quadratics behave very like the quadratics in one variable familiar from school mathematics. They are clearly degenerate: the matrix of a binary quadratic has a row (and column) of zeros, so has a zero determinant. You can view a binary quadratic as a conic for which the origin is a singular point. The point of the next example is that in fact all singular conics are degenerate: the argument uses a consequence of the Invariance Theorem in Chapter 14, namely that the relation $\Delta = 0$ is invariant under translations.

Example 6.1 Any singular conic Q is degenerate. Since Q is singular it has a centre W lying on Q. Translating W to the origin we obtain a translate R of Q for which the origin is a singular point. Thus R is a binary quadratic, hence degenerate. Since the relation $\Delta = 0$ is invariant under translations we conclude that Q likewise is degenerate.

The point of the next result is that the sign of the delta invariant determines whether a binary quadratic is reducible or irreducible.

Lemma 6.1 *Let Q be a binary quadratic defined by (6.1). When $\delta < 0$ it is a real line-pair with vertex the origin; when $\delta = 0$ it is a repeated line through the origin; and when $\delta > 0$ it is irreducible.*

Proof Assume first that one of a, b is non-zero. Then we have the identities

$$a Q(x, y) = (ax + hy)^2 + \delta y^2, \qquad b Q(x, y) = \delta x^2 + (hx + by)^2.$$

When $\delta < 0$ these identities show that Q is a difference of squares, hence a pair of lines through the origin. When $\delta = 0$ they express Q explicitly as a repeated line through the origin. And when $\delta > 0$ they show that the zero set of Q is the origin, so Q is irreducible. (As we pointed out previously, the zero set of a reducible conic is infinite.) It remains to consider the case when $a = b = 0$: but then $\delta < 0$, and Q evidently reduces to two distinct lines through the origin. $\qquad\square$

Example 6.2 The binary quadratic $Q = x^2 + 2hxy + y^2$ with $h > 0$ has delta invariant $\delta = 1 - h^2$. For $h > 1$ we have $\delta < 0$ and Q is a real line-pair. For $h = 1$ it is a repeated line $(x + y)^2$. And for $0 < h < 1$ we have $\delta > 0$ so the origin is the only point in the zero set. An interesting consequence is that distinct values h_1, h_2 of h in the range $0 < h < 1$ give rise to different conics having the same zero set. Unlike the virtual circles of Example 3.2 the zero sets here are non-empty.

Given that the binary quadratic (6.1) has a negative delta invariant we can factorize it on the basis of the following observation. When $a = 0$ or $b = 0$ the factorization is clear, so we can assume $a, b \neq 0$. Write λ, μ for the roots of the quadratic equation $az^2 + 2hz + b = 0$. Then, writing $z = x/y$, and clearing denominators, we see that

$$Q(x, y) = a(x - \lambda y)(x - \mu y).$$

Example 6.3 The binary quadratic $Q = 77x^2 + 78xy - 27y^2$ is reducible as $\delta < 0$: indeed $Q = (11x - 3y)(7x + 9y)$.

6.2 Reducible Conics

In this section we will show that reducible conics are automatically degenerate, and that within that class real line-pairs are characterized by the sign of their delta invariant.

Lemma 6.2 *Any reducible conic Q is degenerate. Moreover, $\delta \le 0$ with equality if and only if Q is a real parallel line-pair.*

Proof Write $L = ax + by + c$, $L' = a'x + b'y + c'$ for the components. The matrix A of the conic $Q = LL'$ is easily checked to be

$$2A = \begin{pmatrix} 2aa' & ab' + a'b & ac' + a'c \\ ab' + a'b & 2bb' & bc' + b'c \\ ac' + a'c & bc' + b'c & 2cc' \end{pmatrix}.$$

Write $v = (a, b, c)$, $v' = (a', b', c')$. The rows of the matrix are $av + a'v'$, $bv + b'v'$, $cv + c'v'$ so the row space is spanned by just the two vectors v, v'. It follows from linear algebra that the determinant Δ is zero. A calculation shows that $\delta = -(ab' - a'b)^2 \le 0$: thus $\delta = 0$ if and only if $ab' - a'b = 0$, which is equivalent to saying that the lines L, L' have the same direction. $\square$

Throughout this text we will tacitly use the logical converse of the first statement in Lemma 6.2, namely that non-degenerate conics are irreducible. That has a useful consequence. According to the Component Lemma any irreducible conic Q intersects a line in at most two points, so non-degenerate conics have the same property. In particular, parabolas, real ellipses, and hyperbolas all meet lines in at most two points. Here is a partial converse of the last result. It uses the fact that any translate of a reducible conic is likewise reducible. (Exercise 6.2.1.)

Lemma 6.3 *Any degenerate conic Q with $\delta < 0$ is reducible.*

Table 6.1. *Degenerate classes*

name	δ	Δ
real line-pairs	$\delta < 0$	$\Delta = 0$
parallel line-pairs	$\delta = 0$	$\Delta = 0$
virtual line-pairs	$\delta > 0$	$\Delta = 0$

Proof Since $\delta < 0$ the conic Q has a unique centre. Translating the centre to the origin we obtain a conic Q' with the same quadratic part, but no terms in x, y. By Lemma 5.5 the constant term in Q' is Δ/δ; since Q is degenerate $\Delta = 0$, so the constant term is likewise zero. Thus Q' is a binary quadratic with $\delta < 0$. It follows from Lemma 6.1 that Q' is reducible, and hence its translate Q is likewise reducible. □

The key consequence of these results is that real line-pairs are characterized as conics for which $\delta < 0$ and $\Delta = 0$. On this basis it makes sense to emulate the non-degenerate case by splitting the degenerate conics into the three classes defined by Table 6.1. It will be convenient to subdivide the 'parallel line-pair' class into three subclasses, namely the real parallel line-pairs, the repeated lines, and the *virtual* parallel line-pairs, characterized by empty zero sets.

It is useful to know in practice whether a given conic Q is reducible, and if so to have a procedure for determining its components. The above generalities suggest how to proceed. A necessary condition for Q to be reducible is that $\Delta = 0$. And in that case, Q is reducible when $\delta < 0$, and irreducible when $\delta > 0$. Of course when $\delta = 0$ the question is undecided. For instance both the conics $Q = x^2 \pm 1$ are degenerate with $\delta = 0$: however, in the '$-$' case Q is a real line-pair, whilst in the '$+$' case Q is irreducible, since its zero set is empty. Even if we know in principle that Q is reducible, we are still faced with the practical question of determining its components. Here is a practical approach. Suppose Q is degenerate, with $\delta \leq 0$. Were it reducible, it could be written in the form

$$Q(x, y) = (px + qy + \alpha)(rx + sy + \beta). \tag{6.2}$$

Observe that the quadratic part of Q would then factorize as the product of the lines $px + qy$, $rx + sy$. That suggests a practical approach: find the factors $px + qy$, $rx + sy$ of the quadratic terms: then determine whether there exist constants α, β for which (6.3) holds, by equating the coefficients of x, y and the constant term. That produces two linear equations in α, β, plus the relation $\alpha\beta = c$. And Q is reducible if and only if all three equations have a solution.

Example 6.4 By way of illustration consider the conic Q defined by

$$Q(x, y) = 3x^2 + 2xy - y^2 - 4x - 4y - 4.$$

The quadratic part is $3x^2 + 2xy - y^2$, with $\delta = -4 < 0$, so reducible: indeed, it factorizes as $(3x + y)(x - y)$. To see whether Q reduces we seek constants α, β for which

$$Q(x, y) = (3x + y + \alpha)(x - y + \beta).$$

Equating coefficients of x, y and the constant term we get $\alpha + 3\beta = -4$, $-\alpha + \beta = -4$, $\alpha\beta = -4$. The first two equations give $\alpha = 2$, $\beta = -2$, which also satisfy the third relation. Thus Q is reducible, indeed

$$Q(x, y) = (3x + y + 2)(x - y - 2).$$

Example 6.5 The trace invariant τ of a general conic Q has received virtually no mention since its introduction in Section 4.3. Recall that $\tau = a + b$, where a, b are the coefficients of x^2, y^2 in Q. In the case when Q is *reducible*, the geometric meaning of the relation $\tau = 0$ is easily ascertained. Any such conic Q can be written in the following form

$$Q(x, y) = (px + qy + \alpha)(rx + sy + \beta). \tag{6.3}$$

The coefficients of x^2, y^2 are $a = pr$, $b = qs$. Thus $\tau = a + b = 0$ is equivalent to the following relation, which holds if and only if the lines are perpendicular

$$0 = pr + qs = (p, q) \bullet (r, s).$$

Exercises

6.2.1 Show that any translate of a reducible conic is itself reducible.

6.2.2 Show that any point circle is in the class of virtual line-pairs.

6.2.3 Let L be a line, and k a constant. Show that the conic $Q = L^2 + k$ is a parallel line-pair, indeed that it is a real parallel line-pair when $k < 0$, a repeated line when $k = 0$, and a virtual parallel line-pair when $k > 0$.

6.2.4 In each of the following cases decide whether the given conic is reducible, and if so find the components:

    (i) $2x^2 + xy - y^2 + x - 2y - 1 = 0$,

    (ii) $x^2 - y^2 + 2\sqrt{2}y - 2 = 0$,

    (iii) $10xy + 8x - 15y - 12 = 0$.

6.2.5   In each of the following cases decide whether the given conic is reducible, and if so find the components:

(i) $2x^2 - xy + 5x - 2y + 2 = 0$,
(ii) $3x^2 + 5xy - 2y^2 - 8x + 5y - 3 = 0$,
(iii) $x^2 - xy + y^2 - x - y + 1 = 0$.

6.2.6   In each of the following cases find the values of the constant λ for which the conic is reducible:

(i) $x^2 + xy + 3x + \lambda y = 0$,
(ii) $2x^2 + 9xy + 4y^2 - \lambda x + 2y = 0$,
(iii) $\lambda xy + 5x + 3y + 2 = 0$.

6.3 Pencils of Conics

The motivation for devoting this chapter to degenerate conics was that they occur naturally as transitional types in *families* of general conics. It is time to amplify this statement. We have already seen how pencils of lines and circles provide natural geometric families. The same is true of general conics. By the *pencil of conics* spanned by two distinct conics Q, R we mean the set of all conics of the form $\lambda Q + \mu R$, where λ, μ are constants, not both zero. Here again we have the same key *intersection property* that we noted for lines and circles, namely that any two distinct elements Q', R' in the pencil have the same intersection as Q, R: the proof is identical to the line case. There *may* be an exceptional ratio $\lambda : \mu$ for which $\lambda Q + \mu R$ fails to be a conic; but for all other ratios it will be a conic. A natural question to ask is how the type of the conic $\lambda Q + \mu R$ depends on the ratio $\lambda : \mu$. We will content ourselves with an example.

Example 6.6 We will investigate the conic types which appear in the pencil of conics $U = \lambda Q + \mu R$, where Q, R are the conics defined by the following formulas

$$Q = 2(y^2 + 2xy + x + y + 1), \qquad R = 2(x^2 - x - y - 1).$$

The reader will readily verify that the delta invariant and discriminant of U are given by

$$\delta = 4\lambda(\mu - \lambda), \qquad \Delta = -2(\lambda - \mu)^2(3\lambda + \mu).$$

The conic U is degenerate if and only if $\Delta = 0$, i.e. $\mu = \lambda$ or $\mu = -3\lambda$. Think of these conditions defining lines through the origin in the (λ, μ)-plane

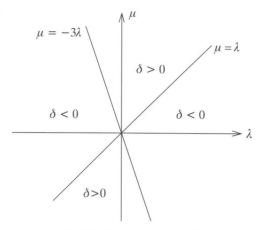

Fig. 6.1. Conic types in a pencil

(Figure 6.1.). The line $\mu = \lambda$ corresponds to the repeated line $S = (x + y)^2$, whilst the line $\mu = -3\lambda$ corresponds to the conic of Example 6.4, namely the real line-pair

$$T = 3x^2 + 2xy - y^2 - 4x - 4y - 4 = -(3x + y + 2)(x - y - 2).$$

Thus our pencil contains two degenerate conics, a repeated line S, and a real line-pair T. Let us pursue the analysis a little further. The lines in the (λ, μ)-plane split it into two cones, any point of which corresponds to a non-degenerate conic in the pencil. The delta invariant gives further information. The conic is a parabola if and only if $\delta = 0$, i.e. $\lambda = 0$ or $\mu = \lambda$. The line $\lambda = 0$ is the μ-axis, and corresponds to our original conic R, which is indeed a parabola. And the line $\mu = \lambda$, as we just seen, corresponds to the repeated line S. The lines $\mu = \lambda$, $\mu = -3\lambda$ determine two cones, one containing the λ-axis, and the other the μ-axis. In the former cone we have $\delta < 0$, so the conic U is a hyperbola. And in the latter, except on the μ-axis itself, we have $\delta > 0$ so U is an ellipse. (In fact, as we will see below, it is a real ellipse.)

In the above example the degenerate conics represent transitional types between ellipses and hyperbolas. The fact that the pencil in this example contains a repeated line is not without significance. In fact it is exceptional for a pencil of conics to contain a repeated line: normally, the degenerate conics in a pencil are line-pairs. In Chapter 8 we will see that the process of finding repeated lines in a pencil of conics is a key step in getting to grips with the geometry of a conic.

Example 6.7 It is worth pursuing the previous example one step further. The intersection property of pencils tells us that the conics Q, R have the same intersections as the degenerate conics S, T, and these are readily verified to be the points $(1, -1)$, $(-1, 1)$. In particular, any conic in the pencil passes through these two points, so has a non-empty zero set. That gives us useful extra information, namely that when U is an ellipse it is actually a *real* ellipse.

What is interesting about this example is that it suggests a strategy for finding the intersections of two quite general conics Q, R by first finding the degenerate conics in the resulting pencil $\lambda Q + \mu R$. That observation turns out to have useful applications within algebra.

Exercise

6.3.1   Consider the pencil of conics $\lambda Q + \mu R$ where Q, R are the conics defined by the following formulas

$$Q = 2(xy - 2x + y), \qquad R = 2(xy - 3x + 2y).$$

Show that the pencil contains three degenerate conics, all of which are real line-pairs. In each of the three cases find the components of the line-pair. Use your results to find the intersections of the conics Q, R. What type of conic is $\lambda Q + \mu R$ when it is non-degenerate?

6.4 Perpendicular Bisectors

In the next chapter we will introduce the general idea of an 'axis' for a conic Q. The purpose of this section is to provide a foretaste of this general concept for the special case when Q be a real line-pair. Suppose Q has components L, M, and vertex P. A *bisector* of Q is a line B through P whose intersection Z with any line perpendicular to B is the midpoint of its intersections with L, M. The mental picture is illustrated in Figure 6.2. The figure suggests that L, M will have two bisectors, and that they will be perpendicular. The function of the next result is to formalize these intuitions, and relate them to the more usual formulation in terms of angles.

Lemma 6.4 *Let L, M be distinct lines in canonical form through a point P. Their bisectors are the lines $L \pm M$, each of which 'bisects' an angle between L, M.*

Proof Let the canonical forms be $L(Z) = U \bullet Z + c$, $M(Z) = V \bullet Z + d$, where U, V are unit vectors. By Example 1.7 any line through P has the form

Degenerate Conics

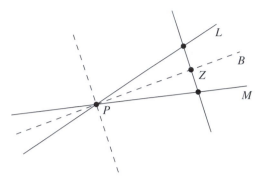

Fig. 6.2. Bisectors of two lines

$B = \alpha L + \beta M$ for some constants α, β. We will determine the condition on α, β for B to be a bisector. Let $Z \neq P$ be a point on B, so that

$$0 = B(Z) = \alpha L(Z) + \beta M(Z). \tag{6.4}$$

The line perpendicular to B through Z has direction $W = \alpha U + \beta V$, so is parametrized as $Z + tW$. The condition for Z to be the midpoint of the intersections with L, M is that there is a non-zero constant t for which $L(Z + tW) = 0$, $M(Z - tW) = 0$. Using the relations $U \bullet U = 1$, $V \bullet V = 1$ that gives

$$\begin{cases} 0 = U \bullet (Z + tW) + c = L(Z) + t\{\alpha + \beta(U \bullet V)\} \\ 0 = V \bullet (Z - tW) + d = M(Z) - t\{\alpha(U \bullet V) + \beta\}. \end{cases}$$

Substituting for $L(Z)$, $M(Z)$ in (6.4) we obtain $\alpha^2 - \beta^2 = 0$, yielding $\beta = \pm \alpha$. It follows that there are just two bisectors, namely the lines $L - M$, $L + M$ with perpendicular directions $U - V$, $U + V$. We claim that the angles α between L, $L - M$ coincide with the angles β between M, $L - M$. We have only to observe that the angles are determined by the following relations, whose right-hand sides are equal

$$\pm \cos \alpha = \frac{U \bullet (U + V)}{|U + V|}, \qquad \pm \cos \beta = \frac{V \bullet (U + V)}{|U + V|}.$$

In exactly the same way we deduce that the angles between L, $L + M$ coincide with the angles between M, $L + M$. $\qquad \square$

In view of this result we refer to the lines $L - M$, $L + M$ as the *perpendicular bisectors* of L, M.

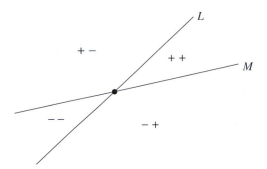

Fig. 6.3. Cones associated to a line-pair

Example 6.8 We will find the perpendicular bisectors of the lines $3x + 2y + 2 = 0$, $18x - y - 1 = 0$. They have canonical forms

$$L(x, y) = \frac{3x + 2y + 2}{\sqrt{13}}, \qquad M(x, y) = \frac{18x - y - 1}{5\sqrt{13}}.$$

It is now a matter of arithmetic to check that the perpendicular bisectors $L + M$, $L - M$ are the lines $11x + 3y + 3 = 0$, $3x - 11y - 11 = 0$.

There is a generality here worthy of comment, since it will be relevant to the geometry of hyperbolas in Chapter 12. The mental picture for a line-pair $Q = LM$ in Figure 6.3 is worth commenting on. For specific choices of linear functions L, M the product Q is a quadratic function, whose sign plays a role. Any linear function divides the plane into two half-planes, one where the function is positive, and the other where it is negative. Thus L, M divides the plane into four *half-cones*, corresponding to the four possible sign pairs $++$, $+-$, $-+$, $--$.

There are therefore two *cones*, with Q positive on one and negative on the other. When we think of L, M as lines, and Q as a line-pair, all that is important is that Q takes *opposite* signs in the two cones. The relevance of these remarks to bisectors is that one bisector of L, M lies in one cone, whilst the other lies in the other cone. We have only to observe that at points (x, y) on $L - M$ we have $LM(x, y) = L^2(x, y)$, and at points on $L + M$ we have $LM(x, y) = -L^2(x, y)$: thus LM takes different signs on the two cones.

Exercises

6.4.1   In each of the following cases find the bisectors of the given pair of lines:

(i) $x + 2y + 3 = 0,$ $2x - y + 3 = 0,$
(ii) $4x + 3y + 10 = 0,$ $12x - 5y + 2 = 0,$
(iii) $x + y + 2 = 0,$ $x - 7y - 2 = 0,$
(iv) $6x + 8y + 13 = 0,$ $2y + 1 = 0.$

6.4.2   Show that the points $(1, 1)$, $(2, 3)$, $(0, 7)$, $(-2, 4)$ lie in the four sectors of the plane determined by the linear functions $3x + 2y - 6$ and $2x - y + 2$.

7

Axes and Asymptotes

A general philosophy of the subject is that it is fruitful to understand how a conic Q intersects lines. It is not just the intersections of Q with a single line which are significant, but its intersections with pencils of lines. In this chapter we pursue this philosophy in the special case of a parallel pencil of lines. Any line in the pencil intersecting Q twice determines a chord with a unique midpoint. Section 7.1 establishes that the midpoints lie on a line, at least provided the delta invariant is non-zero. In the next section we use this 'midpoint locus' to introduce axes of symmetry, a significant visual feature of a conic. For instance ellipses and hyperbolas have two perpendicular axes through their centre, whilst a parabola has a single axis. Moreover, for line-pairs the axes are the perpendicular bisectors of the component lines, familiar from Chapter 6. The final section illustrates an exceptional situation, namely parallel pencils whose *general* line meets Q in a single point: that leads us to the concept of 'asymptotic directions' for a conic, and the associated classical idea of an 'asymptote'.

7.1 Midpoint Loci

Consider the intersection of a conic Q with the parallel pencil of lines in a general direction (X, Y). This section revolves around the fact that the midpoints of the resulting family of parallel chords lie on a line, the 'midpoint locus' associated to the direction (X, Y). For instance, when Q is a real circle, the locus of midpoints is a diameter. Figure 7.1 illustrates a less familiar example, when Q is an ellipse.

We have to be careful when formulating a definition, since the lines in the pencil may not intersect Q: for instance, the zero set of Q may be empty. To provide motivation, we write down a condition we expect to be satisfied on the potential midpoints $Z = (x, y)$. Consider the line L through Z in the direction

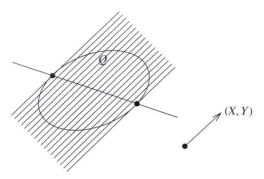

Fig. 7.1. A midpoint locus for an ellipse

(X, Y), parametrized as

$$x(t) = x + tX, \qquad y(t) = y + tY.$$

The point Z itself corresponds to the parameter $t = 0$. The idea is that for Z to be a midpoint of a chord we expect two values $t = t_0, t = -t_0$ for which the point with parameter t lies on Q. In Section 4.4 we showed that the intersections of L with Q are determined by the roots of a quadratic equation

$$\phi(t) = pt^2 + qt + r = 0.$$

The condition for the roots to be equal and opposite is that their sum is zero, i.e. $q(X, Y) = 0$. We need to interpret this for a general conic

$$Q(x, y) = ax^2 + 2hxy + by^2 + 2gx + 2fy + c. \qquad (\star)$$

Using the relations (4.3) for the coefficients p, q, r in the quadratic, we see that the required condition is that

$$(ax + by + g)X + (hx + by + f)Y = 0. \qquad (7.1)$$

We are thinking of the direction (X, Y) as fixed, and x, y as the variables, so it is more illuminating to write this in the form

$$(aX + hY)x + (hX + bY)y + (gX + fY) = 0. \qquad (7.2)$$

The set of points (x, y) for which this relation holds is the *midpoint locus* for the direction (X, Y): it is unchanged when we replace (X, Y) by a non-zero scalar multiple. The key facts are highlighted by the following statement.

Lemma 7.1 *The midpoint locus of Q associated to the direction (X, Y) is either a line, or empty, or the whole plane. Moreover, it contains every centre.*

When $\delta \neq 0$ *the midpoint locus is a line (through the unique centre) for any choice of direction* (X, Y).

Proof The first statement is immediate from the form of (7.2). By Lemma 5.3, any centre (u, v) for Q satisfies the equations

$$au + hv + g = 0, \qquad hu + bv + f = 0.$$

Then by (7.1) the point (u, v) lies on the midpoint locus, for any choice of direction. That establishes the second statement of the lemma. Suppose now that $\delta \neq 0$. The midpoint locus associated to a direction (X, Y) is a line, provided one of the coefficients of x, y in (7.2) is non-zero: that fails only when both coefficients vanish

$$aX + hY = 0, \qquad hX + bY = 0.$$

Since $\delta = ab - h^2 \neq 0$ these linear equations in X, Y only have the trivial solution $X = Y = 0$, a contradiction. Thus at least one of the coefficients of x, y is non-zero, and the midpoint locus is a line. $\qquad\square$

Example 7.1 The real ellipse $x^2 + 2y^2 = 1$ has the origin as its unique centre. The midpoint locus associated to the direction (X, Y) is the diameter $Xx + 2Yy = 0$. For instance, the locus associated to the direction $(1, 1)$ is the line $x + 2y = 0$.

Example 7.2 For a standard parabola $y^2 = 4ax$ with $a > 0$ we have $\delta = 0$, so there is no guarantee that the midpoint locus will be a line for every direction (X, Y). The midpoint locus has equation $2aX - yY = 0$, so provided $Y \neq 0$ it is a horizontal line. However for the direction $(1, 0)$ of the x-axis the locus is defined by $2a = 0$, so is empty.

It is worth remarking that our conclusions depend crucially on the fact that we have considered the intersections of a conic Q with a *parallel* pencil of lines. The next example illustrates what can happen when we take a *general* pencil.

Example 7.3 Let Q be the standard parabola $y^2 = 4ax$. Consider the intersections of Q with the pencil of lines through the origin. Only one line in the pencil meets Q just once, namely $y = 0$. Any other line has the form $x = ty$ for some constant t, and meets Q at the origin and the point $(4at^2, 4at)$. The midpoint of the chord through these points is $(2at^2, 2at)$ so lies on a second parabola $y^2 = 2ax$. Thus the locus of midpoints fails to be a line.

Exercises

7.1.1   In each of the following cases find the midpoint locus associated to the given conic Q, and the given direction $Z = (X, Y)$:

(i) $Q = ax^2 + by^2$, $Z = (1, 1)$,
(ii) $Q = xy - 1$, $Z = (X, Y)$,
(iii) $Q = 5x^2 - 2xy + 4x$, $Z = (1, -1)$.

7.1.2   Show that there is a unique direction (X, Y) for which the midpoint locus of the real parallel lines $y^2 = 1$ fails to be a line. Verify that for *any* direction it contains all the centres.

7.2 Axes

The standard ellipses and hyperbolas of Section 4.1 had evident symmetry in both coordinate axes: likewise, the standard parabolas had evident symmetry in the x-axis. Our objective in this section is to introduce 'axes of symmetry' for general conics. It turns out that any conic is symmetric in at least one line, possibly more: for instance, a circle is symmetric about *every* line through its centre. Such lines are fundamental to the geometry of the conic, and we need practical techniques for finding them. A line L is an *axis* of a conic Q when it is the midpoint locus associated to a perpendicular direction; and points where an axis L meets Q are *vertices* of Q. Observe that an axis L is necessarily a line of symmetry, in the sense that the line through any point P on the conic perpendicular to L meets Q again at a point P' equidistant from L.

Example 7.4 Any line through the centre of the real circle $x^2 + y^2 = r^2$ with $r > 0$ is an axis. The relation (7.2) tells us that the midpoint locus associated to any direction (X, Y) is the line $Xx + Yy = 0$ through the centre. The claim follows from the observation that *any* line through the centre has this form, so is an axis. It follows that every point on the circle is a vertex.

   Our immediate objective is to find axes (and hence vertices) in practice. As always, we consider a general conic

$$Q(x, y) = ax^2 + 2hxy + by^2 + 2gx + 2fy + c. \qquad (\star)$$

Lemma 7.2 *The midpoint locus associated to a non-zero vector (X, Y) is an axis for the conic $(\star)$ if and only if there exists a non-zero constant λ for which*

$$aX + hY = \lambda X, \qquad hX + bY = \lambda Y. \qquad (7.3)$$

Proof The condition for the midpoint locus to be an axis is that some line $Xx + Yy + Z$ perpendicular to the direction (X, Y) coincides with the associated midpoint locus (7.2). That is equivalent to saying that there is a non-zero constant λ for which

$$aX + hY = \lambda X, \qquad hX + bY = \lambda Y, \qquad gX + fY = \lambda Z.$$

In particular (7.3) holds for some $\lambda \neq 0$. Conversely, if (7.3) holds for some non-zero constant λ then we can *define* Z by $gX + fY = \lambda Z$, and conclude that the line $Xx + Yy + Z$ coincides with the midpoint locus, so is an axis.

$\square$

Let us look more closely at the relations (7.3). It will be helpful to write them as a single matrix relation

$$\begin{pmatrix} a & h \\ h & b \end{pmatrix} \begin{pmatrix} X \\ Y \end{pmatrix} = \lambda \begin{pmatrix} X \\ Y \end{pmatrix}. \tag{7.4}$$

Any real number λ for which this holds for some non-zero vector (X, Y) is a *eigenvalue* of the quadratic function $(\star)$. We stress that λ is associated to $(\star)$, not to the conic it defines: the result of multiplying Q by a constant k is to multiply λ by k. Given an eigenvalue λ, any vector (X, Y) for which (7.4) holds is a *eigenvector* associated to that eigenvalue. Note that if (X, Y) is a eigenvector associated to λ then any scalar multiple is also a eigenvector. To find the eigenvalues, write (7.4) in the form

$$\begin{pmatrix} a - \lambda & h \\ h & b - \lambda \end{pmatrix} \begin{pmatrix} X \\ Y \end{pmatrix} = \begin{pmatrix} 0 \\ 0 \end{pmatrix}. \tag{7.5}$$

Then λ is an eigenvalue if and only if this holds for some non-zero vector (X, Y). By linear algebra, that is equivalent to

$$\begin{vmatrix} a - \lambda & h \\ h & b - \lambda \end{vmatrix} = 0. \tag{7.6}$$

Expansion of the determinant yields the following quadratic in λ, known as the the *characteristic equation*, whose roots are the required eigenvalues

$$\lambda^2 - (a + b)\lambda + (ab - h^2) = 0. \tag{7.7}$$

The characteristic equation always has at least one real root, since its discriminant is $(a - b)^2 + 4h^2 \geq 0$. It has a repeated root if and only $a = b$ and $h = 0$, so precisely when $(\star)$ is a circle. With that sole exception, there are always two distinct eigenvalues, with product the delta invariant $\delta = ab - h^2$. For an ellipse the eigenvalues have the same sign, whilst for the hyperbola they have different signs. Note that there is a zero eigenvalue if and only if the constant term $\delta = 0$: however, zero cannot be a repeated root as the relations $a = b$, $h = 0$, $\delta = 0$ hold if and only if $a = b = h = 0$. Although the magnitude and sign of an eigenvalue may well change on multiplying ($\star$) by a constant, it does make sense to refer to the eigenvalue of smaller or larger absolute value.

It follows from the above that a direction (X, Y) perpendicular to an axis is necessarily an eigenvector associated to *non-zero* eigenvalues λ. Incidentally, since at least one eigenvalue is non-zero we have established that *every conic has at least one axis*. In the case when $\delta \neq 0$, and ($\star$) is not a circle, there are two non-zero eigenvalues giving rise to two axis directions (X, Y). The corresponding axes are the lines $Xx + Yy + Z = 0$ through the unique centre.

Example 7.5 Consider any conic Q given by a formula of the form $Q = px^2 + qy^2 - 1$, where p, q are distinct non-zero constants. Clearly, Q has the origin as its unique centre. The characteristic equation is displayed below, yielding eigenvalues $\lambda = p, q$

$$\lambda^2 - (p + q)\lambda + pq = 0.$$

Associated eigenvectors are the directions $(1, 0)$, $(0, 1)$ of the x- and y-axes. Morover the coordinate axes pass through the centre, so are the required axes of Q. In particular the axes of the standard ellipses and hyperbolas are the coordinate axes, consistent with the definitions given in Section 4.1.

Example 7.6 The reader will readily verify that the displayed conic Q is an ellipse, with centre $(-1, 1)$

$$Q(x, y) = 17x^2 - 12xy + 8y^2 + 46x - 28y + 17.$$

The characteristic equation is $\lambda^2 - 25\lambda + 100 = 0$ with roots $\lambda = 5$, $\lambda = 20$. Corresponding eigenvectors are then given by

$$\begin{pmatrix} 12 & -6 \\ -6 & 3 \end{pmatrix} \begin{pmatrix} X \\ Y \end{pmatrix} = \begin{pmatrix} 0 \\ 0 \end{pmatrix}, \qquad \begin{pmatrix} -3 & -6 \\ -6 & -12 \end{pmatrix} \begin{pmatrix} X \\ Y \end{pmatrix} = \begin{pmatrix} 0 \\ 0 \end{pmatrix}.$$

That yields respective eigenvectors $(1, 2)$, $(2, -1)$. The axes are the lines through the centre in these directions, namely

$$2x - y + 3 = 0, \qquad x + 2y - 1 = 0.$$

It is no coincidence that the eigenvectors in the above examples are perpendicular: that is a consequence of the following generality.

Lemma 7.3 *Any direction* (X, Y) *perpendicular to an axis for* $(\star)$ *satisfies the binary quadratic equation*

$$hX^2 + (b - a)XY - hY^2 = 0. \tag{7.8}$$

Proof Another way of expressing the conditions (7.3) is to say that the vectors (X, Y), $(aX + hY, hX + bY)$ are linearly dependent. By linear algebra that is equivalent to the vanishing of their determinans. And that is precisely (7.8).
□

We call (7.8) the *direction quadratic* for $(\star)$. The coefficients vanish when $a = b$, $h = 0$, so if and only if there is a repeated eigenvalue. Otherwise, the direction quadratic has a positive discriminant, so determines two directions (X, Y). Moreover, we can deduce from Example 6.5 that these directions are perpendicular, since the sum of the coefficients of X^2, Y^2 is zero. The direction quadratic offers another way of finding axis directions.

Example 7.7 In the previous example the coefficients of the quadratic terms were $a = 17$, $b = 8$, $h = -6$ so the direction quadratic is as follows, yielding the same axis directions

$$0 = -3(2X^2 + 3XY - 2Y^2) = -3(2X - Y)(X + 2Y).$$

Example 7.8 For the standard parabola $Q = y^2 - 4ax$ with $a > 0$ the characteristic equation is $\lambda(\lambda - 1) = 0$. Thus the eigenvalues are $\lambda = 0, 1$ with respective eigenvectors the directions $(1, 0)$, $(0, 1)$ of the coordinate axes. In Example 7.2 we saw that the midpoint locus for the y-axis direction is the line $y = 0$, whilst that for the x-axis direction is empty. There is therefore just one axis $y = 0$ corresponding to the non-zero eigenvalue.

Example 7.9 The geometric condition for a general conic $(\star)$ to have an axis direction (X, Y) parallel to a coordinate axis is that (7.8) should have a solution with $X = 0$ or $Y = 0$. Clearly, that is equivalent to the algebraic condition $h = 0$. The initial step in classifying conics in Chapter 15 will be to force the algebraic condition $h = 0$ by applying a suitable 'rotation' of the conic about the origin.

Exercises

7.2.1   In each of the following cases verify that the given conic is an ellipse, and find its centre and axes:

(i) $x^2 - xy + 2y^2 = 0$,

(ii) $5x^2 - 24xy - 5y^2 + 14x + 8y - 16 = 0$,

(iii) $3x^2 + 2xy + 3y^2 + 14x + 20y - 183 = 0$.

7.2.2   In each of the following cases find the centre and axes of the given conic:

(i) $x^2 + xy + y^2 - 2x + 2y - 6 = 0$,

(ii) $3x^2 + 2xy + 3y^2 - 6x + 14y - 101 = 0$,

(iii) $77x^2 + 78xy - 27y^2 + 70x - 30y + 29 = 0$.

7.3 Bisectors as Axes

A special case of the axis construction arises when Q is a real line-pair, with components L, L' and vertex P. In that case $\delta < 0$, so, by the above, discussion there are two perpendicular axes M, M'. Clearly, these are the perpendicular bisectors of Section 6.4. The simplest case is when L, L' are lines through the origin, with joint equation

$$Q(x, y) = ax^2 + 2hxy + by^2.$$

Then, by the above, the joint equation of the bisectors is the following binary quadratic; factorization yields the individual bisectors

$$B(x, y) = hx^2 + (b - a)xy - hy^2.$$

A potential advantage of this approach over the method described in Section 6.4 is that starting from the joint equation of the lines, one proceeds directly to the joint equation of the bisectors. However, there is the potential drawback of having to factorize the joint equation of the bisectors.

Example 7.10 The joint equation of the lines $y = 2x$, $y = 3x$ through the origin is $6x^2 - 5xy + y^2$. By the above, the joint equation of the bisectors is $5(x^2 + 2xy - y^2)$. Factorizing, we see that the bisectors are the lines

$$x + (1 - \sqrt{2})y = 0, \qquad x + (1 + \sqrt{2})y = 0.$$

The procedure can be modified for line-pairs $Q(x, y)$ whose centre is not the origin. Find the centre (u, v): then the translation $x = X + u$, $y = Y + v$ gives a pair of lines $R(X, Y)$ through the origin. The joint equation $R'(X, Y)$ of its bisectors can then be read-off as above. Finally, translating back by $X = x - u$, $Y = y - v$ we obtain the joint equation $Q'(x, y)$ of the bisectors for Q.

Example 7.11 We leave the reader to verify that the conic Q below is a line-pair with centre $(-3, 1)$

$$Q(x, y) = x^2 - 5xy + 3y^2 + 11x - 21y + 27.$$

The translation $x = X - 3$, $y = Y + 1$ taking the centre to the origin then yields the parallel line-pair $R(X, Y)$ below. The joint equation of the perpendicular bisectors is then a multiple of $R'(X, Y)$

$$R(X, Y) = X^2 - 5XY + 3Y^2, \qquad R'(X, Y) = 5X^2 - 4XY - 5Y^2.$$

Translating back to the centre of Q, by setting $X = x + 3$, $Y = y - 1$ we see that its perpendicular bisectors have joint equation

$$Q'(x, y) = 5x^2 - 4xy - 5y^2 + 34x - 2y + 52.$$

Exercises

7.3.1   Use the method of Section 6.4 to find the bisectors of the lines $y = 2x$, $y = 3x$. Verify that your answer agrees with that obtained in Example 7.10.

7.3.2   Show that the joint equation of the bisectors for the distinct lines $y = \lambda x$, $y = \mu x$ is

$$(\lambda + \mu)x^2 + 2(\lambda\mu - 1)xy - (\lambda + \mu)y^2 = 0.$$

7.3.3   Let $ax^2 + 2hxy - ay^2 = 0$ be the joint equation of the bisectors of a line-pair. Show that the line-pair must have the following form for some constant α

$$(\alpha - h)x^2 + 2axy + (\alpha + h)y^2 = 0.$$

7.3.4   In each of the following cases find the perpendicular bisectors of the given line-pairs:

(i) $3x^2 + 4xy + 2y^2$,
(ii) $2x^2 - 3xy + y^2$,
(iii) $x^2 - 5xy + 3y^2 + 11x - 21y + 27$.

7.3.5   Let L, L' be parallel lines. Show that their parallel bisector is an axis of the real parallel line-pair LL'.

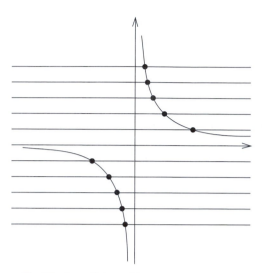

Fig. 7.2. A parallel pencil intersecting a hyperbola

7.4 Asymptotic Directions

For a given conic Q, each direction in the plane determines a parallel pencil
of lines, giving rise to an associated midpoint locus. Normally the locus is a
line, but exceptionally it may be empty, or even the whole plane. In this section
we turn our attention to these exceptional cases. The motivation is provided by
illustrations of hyperbolas, which suggest that there are directions in which the
curve tends to a 'point at infinity'. For instance as $t \to \infty$ so the point $x = t$,
$y = 1/t$ on the hyperbola $xy = 1$ becomes ever closer to the x-axis. In this
sense the hyperbola has a 'point at infinity' on the x-axis, and likewise one on
the y-axis. The reader can find out more about this idea in *EGAC*, where the
ordinary plane is extended to the 'projective' plane by the addition of idealised
'points at infinity' which need to be taken into account. Thus when considering
the intersections of Q with a parallel pencil the mental picture is that Q meets
the general line of the pencil at two points. For directions in which Q tends to
a 'point at infinity' that mental picture persists, provided we think of every line
in the pencil having just one intersection with Q in the Euclidean plane, plus
another 'at infinity'. Figure 7.2 illustrates the parallel pencil of horizontal lines
intersecting the hyperbola $xy = 1$.

That is the motivation for studying parallel pencils in which every line L
meets a conic Q *at most once*. Consider therefore a line L in a direction (X, Y)
through a point (u, v). The intersection quadratic associated to Q and L has
the following form, where the coefficients are given by the formulas (4.3)

$$\phi(t) = pt^2 + qt + r.$$

Our interest lies in the case when the quadratic has at most one root, motivating the following definition. A direction (X, Y) is *asymptotic* for Q when $p(X, Y) = 0$. We need to interpret that for a general conic $(\star)$. According to (4.3) the coefficient $p(X, Y)$ is zero when

$$p(X, Y) = a^2 X^2 + 2hXY + b^2 Y^2 = 0.$$

That is a binary quadratic equation in X, Y. It has two distinct roots when $\delta < 0$, a repeated root when $\delta = 0$, and no roots when $\delta > 0$: thus Q has two asymptotic directions when $\delta < 0$, just one when $\delta = 0$, and none when $\delta > 0$. In particular hyperbolas have two asymptotic directions, parabolas have one, and ellipses have none.

Example 7.12 For the hyperbola $x^2 + 2xy - 3y^2 + 8y + 1 = 0$ asymptotic directions (X, Y) are given by the displayed binary quadratic equation, with roots $(1, 1)$ and $(3, -1)$

$$0 = X^2 + 2XY - 3Y^2 = (X - Y)(X + 3Y).$$

An *asymptote* of a conic Q is a line L in an asymptotic direction that does not intersect Q. In that case the intersection quadratic has no roots, so *both* coefficients p, q vanish. Ellipses cannot have asymptotes since they do not have asymptotic directions: and in Chapter 10 we shall see that parabolas likewise fail to have asymptotes, despite having a unique asymptotic direction. However in Chapter 12 we will see that every hyperbola has two asymptotes through its centre, and we will describe practical procedures for finding them.

Exercise

7.4.1　In each of the following cases, find asymptotic directions for the given conic:

(i)　$x^2 + 2xy + 3y^2 + 8y + 1 = 0$,
(ii)　$4x^2 - 4xy + y^2 - 10y - 19 = 0$,
(iii)　$x^2 + 2xy - 3y^2 + 8y + 1 = 0$.

8

Focus and Directrix

A key feature of a real circle is that it can be constructed *metrically* as the locus of points at a constant positive distance from a fixed point F. The initial object of this chapter is to present a metric construction that produces parabolas, real ellipses, and hyperbolas. The construction involves a fixed point F, and a fixed line D not passing through F. We will show that the standard parabolas, real ellipses, and hyperbolas of Chapter 4 can all be constructed in this way. Indeed, combining this with the classification of conics in Chapter 15, we will see that *any* parabola, real ellipse, or hyperbola has this property. The importance of the construction lies in the fact that constructible conics have interesting metric properties, of physical significance. That leads us to geometry which might otherwise remain unnoticed.

8.1 Focal Constructions

For the purposes of this text a *focal construction* (or just *construction*) is a choice of a point F (the *focus*), a line D not through F (the *directrix*), and a positive constant e (the *eccentricity*). We consider a variable point P subject to the constraint that its distance from F is proportional to its distance from D, where the constant of proportionality is e. Since distances are automatically non-negative, the constraint can be written in the following form, where PD is the distance from P to D

$$PF^2 = e^2 PD^2. \tag{8.1}$$

Example 8.1 Let a be a positive constant. Consider the construction whose focus is the point $F = (a, 0)$, whose directrix is the line $x = -a$, and whose eccentricity is $e = 1$. Writing $P = (x, y)$ the relation (8.1) becomes the following, which we recognize as a standard parabola

$$(x - a)^2 + y^2 = (x + a)^2.$$

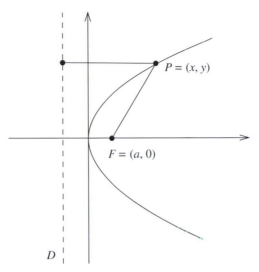

Fig. 8.1. Construction of the standard parabola

We can rewrite the general relation (8.1) explicitly using the formula (2.1) for the distance from a point to a line. To this end let $P = (x, y)$, let $F = (\alpha, \beta)$, and let $D = px + qy + r$ be the equation of the directrix in canonical form. Then (8.1) becomes

$$Q(x, y) = (x - \alpha)^2 + (y - \beta)^2 - e^2(px + qy + r)^2 = 0. \quad (8.2)$$

Example 8.2 Consider the construction with focus $F = (-1, 3)$, directrix $D = 2x - y + 1$, and eccentricity $e = \sqrt{5}$. In this case the reader will readily verify that (8.1) becomes

$$3x^2 - 4xy + 2x + 4y - 9 = 0.$$

Returning to the general case, the coefficients of the quadratic terms x^2, xy, y^2 in (8.2) are displayed below. Clearly they cannot vanish simultaneously, so the formula does actually define a conic Q

$$1 - e^2 p^2, \qquad -2e^2 pq, \qquad 1 - e^2 q^2.$$

The conics Q arising in this way are said to be *constructible*, and (8.2) is a *constructible form* of the equation. By direct calculation we see that the invariants of Q are

$$\tau = 2 - e^2, \qquad \delta = 1 - e^2, \qquad \Delta = -e^2(p\alpha + q\beta + r)^2. \quad (8.3)$$

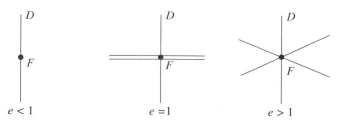

Fig. 8.2. Degenerate 'constructible' conics

We claim that constructible conics are automatically non-degenerate, and hence irreducible. Indeed, the condition $\Delta = 0$ is equivalent to $p\alpha + q\beta + r = 0$, which holds if and only if F lies on D, contrary to assumption. Moreover, Q is an ellipse when $e < 1$, a parabola when $e = 1$, and a hyperbola when $e > 1$.

Example 8.3 Circles are not constructible conics. Indeed (8.2) is a circle if and only if the coefficients of x^2, y^2 are equal, and the coefficient of xy is zero. That is the case if and only if $p = q = 0$, a contradiction.

Example 8.4 Let Q be the conic arising from a construction with focus F and directrix D. Then D cannot intersect Q. Any intersection P satisfies $PD = 0$ so $PF = 0$, contradicting the assumption that F does not lie on D. Likewise, F cannot lie on Q: otherwise, taking $P = F$ we obtain $FD = 0$, another contradiction.

Example 8.5 If in the construction we allow F to lie on D we obtain a degenerate conic Q, indeed a real line-pair with vertex F when $e > 1$, a parallel line-pair when $e = 1$, and a virtual line-pair when $e < 1$. We leave the reader to check that when $e = 1$ the conic Q is actually a repeated line, the line being that through F perpendicular to D.

Exercises

8.1.1   In each of the following cases determine the conic with the given focus F, directrix D, and eccentricity e:

(i) $F = (-1, 1)$, $D = x - y + 3$, $e = 1/2$,
(ii) $F = (3, 0)$, $D = 2x + y + 4$, $e = 1$,
(iii) $F = (-1, 3)$, $D = 2x - y + 1$, $e = \sqrt{5}$.

8.1.2   Determine the conic with centre the origin having focus $(3, 0)$ and directrix the line $x = 1$.

8.1.3   Verify the formulas (8.3) for the invariants of the general constructible conic Q.

8.2 Principles for Finding Constructions

Let Q be a non-degenerate conic. Let us assume that it is constructible, and ask how we might go about determining the focus F, the directrix D, and the eccentricity e associated to that construction. Here are some general principles, leading to a practical technique. We seek a point $F = (\alpha, \beta)$, a line $D = px + qy + r$ in canonical form, and a positive constant e, for which Q is defined by the quadratic function of (8.2)

$$(x - \alpha)^2 + (y - \beta)^2 - e^2(px + qy + r)^2 = 0.$$

For notational efficiency write this as $C - e^2 D^2$, where C is the point circle with centre F defined by

$$C(x, y) = (x - \alpha)^2 + (y - \beta)^2.$$

Our assumption is that $Q = \lambda(C - e^2 D^2)$ for some non-zero scalar λ. It follows that $Q - \lambda C$ is a constant multiple of D^2, hence a repeated line. That suggests how we might proceed. We could take an arbitrary point F, and look for repeated lines in the pencil of conics $Q - \lambda C$. At this point a useful observation is that the delta invariant of a repeated line is zero. Thus we could start by trying to find at least those λ for which the delta invariant of $Q - \lambda C$ is zero. To make this explicit, write Q in the usual form

$$Q(x, y) = ax^2 + 2hxy + by^2 + 2gx + 2fy + c. \qquad (\star)$$

The values of λ for which the delta invariant of $Q - \lambda C$ is zero are easily checked to be solutions of the following quadratic equation

$$\lambda^2 - (a + b)\lambda + (h^2 - ab) = 0.$$

That is precisely the characteristic equation (7.5) of the quadratic function Q. For each eigenvalue λ we can then try to discover whether there are values of α, β for which $Q - \lambda C$ is a repeated line. In the following sections we will turn these ideas into practical techniques.

8.3 Constructions for Parabolas

A special case arises when Q is a parabola, so has zero delta invariant. In that case $\lambda = 0$ must be an eigenvalue, and the other is $\lambda = a + b$. The eigenvalue

$\lambda = 0$ is of no interest to us, since Q itself cannot be a repeated line. The question is whether for the non-zero eigenvalue $\lambda = a + b$ the conic $Q - \lambda C$ is a repeated line, for some choice of α, β.

Lemma 8.1 *The standard parabola with modulus a has a unique construction, with focus $F = (a, 0)$, directrix $D = x + a$, and eccentricity $e = 1$.*

Proof Consider a standard parabola $Q(x, y) = y^2 - 4ax$ with $a > 0$. The characteristic equation is $\lambda(\lambda - 1) = 0$ giving eigenvalues $\lambda = 0, 1$. Moreover

$$Q - C = -x^2 - 2(2a - \alpha)x + 2\beta y - (\alpha^2 + \beta^2).$$

To be a repeated line this needs to have the form $-(px + qy + r)^2$. And its quadratic part $-(px + qy)^2$ would have to be $-x^2$, so we could assume $p = 1$, $q = 0$. Thus we seek a constant r for which

$$Q - C = -(x + r)^2 = -x^2 - 2rx - r^2.$$

That is only possible when $2a - \alpha = r$, $\beta = 0$, $\alpha^2 + \beta^2 = r^2$. Substituting for α, β in the third relation gives a *unique* solution $r = a$ for which $\alpha = a$, $\beta = 0$. In this way we obtain the following constructible form for a standard parabola

$$(x - a)^2 + y^2 = (x + a)^2.$$

$\square$

This is of course the construction of the standard parabola given in Example 8.1. It will follow from this lemma and the listing of conics in Chapter 15, that *any* parabola has a unique focus and directrix, which we can determine in any given example by mimicking the calculation for the standard parabola.

Example 8.6 For the parabola $Q = 4x^2 - 4xy + y^2 - 10y - 19$ the characteristic equation is $\lambda(\lambda - 5) = 0$, giving eigenvalues $\lambda = 0, 5$. The latter gives

$$Q - 5C = -(x + 2y)^2 + 10\alpha x + 10(\beta - 1)y - (5\alpha^2 + 5\beta^2 + 19).$$

We seek α, β for which this conic is a repeated line. A necessary condition is that it should have the same quadratic terms, up to a constant multiple. Thus the above expression needs to have the form $-(x + 2y + r)^2$ for some constant r. Comparing coefficients of x, y and the constant term we see that α, β must satisfy the relations

$$10\alpha = -2r, \qquad 10(\beta - 1) = -4r, \qquad 5\alpha^2 + 5\beta^2 + 19 = r^2.$$

Solving the first two relations for α, β in terms of r and substituting in the third relation gives the unique solution $r = 6$. Thus the focus and directrix are

$$F = \left(-\frac{6}{5}, -\frac{7}{5}\right), \qquad D = x + 2y + 6.$$

Exercise

8.3.1   In each of the following cases verify that the given conic is a parabola and find its axis, vertex, focus, and directrix:

(i) $x^2 + 2xy + y^2 - 3x + 6y - 4 = 0$,
(ii) $x^2 - 4xy + 4y^2 + 10x - 8y + 13 = 0$,
(iii) $4x^2 - 4xy + y^2 + 22x - 6y + 24 = 0$,
(iv) $4x^2 - 4xy + y^2 + 6x - 18y + 36 = 0$.

8.4 Geometric Generalities

Before moving on to the question of finding foci and directrices of real ellipses and hyperbolas it helps to spell out some generalities. The first is that foci always lie on axes.

Lemma 8.2 *Let Q be a conic arising from a construction with focus F and directrix D. Then the line A through F perpendicular to D is an axis for Q.*

Proof We have to show A is the midpoint locus associated to its perpendicular direction. Let L be any line parallel to D, intersecting A at M, and Q at P. (Figure 8.3.) Since constructible conics are irreducible, the Component Lemma ensures that L meets Q at most twice, and we need to show M is their midpoint. Assume P, M are distinct, and let P' be the other point on L equidistant from F. Then $PF = ePD$, $PF = P'F$, $PD = P'D$. It follows that $P'F = eP'D$, so P' is a second intersection of L with Q, and M is the midpoint. When P, M coincide, there is nothing to prove. $\qquad\square$

Thus a focus F of any constructible conic Q lies on an axis A, and its associated directrix D is perpendicular to A. In particular, when Q has two distinct foci F, F' we deduce that the line A joining them is an axis. In fact *any* central constructible conic has two distinct constructions, as we might expect from symmetry considerations. (The ellipse case is illustrated in Figure 8.4, and the hyperbola case in Figure 8.5.) The following lemma verifies this formally. We will refer to the two constructions of the lemma as *mirror* constructions, since each is obtained from the other by central reflexion in the centre.

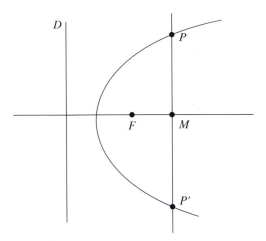

Fig. 8.3. Axis of a constructible conic

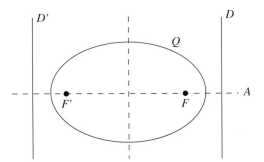

Fig. 8.4. Constructions of a standard ellipse

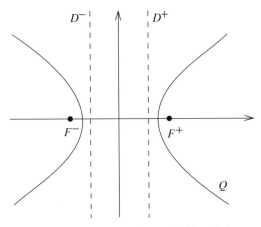

Fig. 8.5. Constructions of a standard hyperbola

Lemma 8.3 *Let Q be a constructible conic with focus F, directrix D, and eccentricity e. Suppose that Q has a unique centre $W = (u, v)$ not on D. Then Q has a second construction with focus F' on the same axis A as F, directrix D' parallel to D, and the same eccentricity e.*

Proof Let $F = (\alpha, \beta)$, let $D = px + qy + r$ be in canonical form. We can suppose that $Q = C - e^2 D^2$, where C is the point circle with centre F. Write $x' = 2u - x$, $y' = 2v - y$ for the central reflexion of the point (x, y) in W. Observe that the conic C' defined below is the point circle with centre the central reflexion $F' = (\alpha', \beta')$ of F in W

$$C'(x, y) = C(x', y') = (x - \alpha')^2 + (y - \beta')^2.$$

Also, the line D' defined below is parallel to D and in canonical form

$$D'(x, y) = D(x', y') = -D(x, y) + 2(pu + qv + r).$$

Since W is a centre for Q, we have the relations

$$\begin{aligned}
Q(x, y) = Q(x', y') \\
= C(x', y') - e^2 D(x', y')^2 \\
= C'(x, y) - e^2 D'(x, y)^2.
\end{aligned}$$

Thus Q has a second construction with focus F', directrix D', and eccentricity e. Since F, F', W are collinear, and the axis A passes through the centre, F' likewise lies on A. Finally, the lines D, D' coincide if and only if $pu + qv + r = 0$, i.e. if and only if D passes through the centre, contrary to hypothesis. $\square$

Exercise

8.4.1   Let F, F' be the foci of a central conic Q. Show that the perpendicular bisector of the line joining F, F' is an axis of Q.

8.5 Constructions of Ellipse and Hyperbola

Let us now extend the philosophy of Section 8.2 to central conics. In that case neither eigenvalue is zero, so we have to consider each one separately. The standard ellipse provides a model for such calculations, simultaneously illustrating Lemma 8.3.

Lemma 8.4 *The standard ellipse with moduli a, b has just two constructions, with eccentricity e defined by $a^2e^2 = a^2 - b^2$. The foci and directrices are*

$$F^+ = (ae, 0), \quad D^+ = x - \frac{a}{e}; \qquad F^- = (-ae, 0), \quad D^- = x + \frac{a}{e}.$$

Proof The standard ellipse with moduli a, b satisfying $0 < b < a$ is defined by the quadratic function

$$Q(x, y) = \frac{x^2}{a^2} + \frac{y^2}{b^2} - 1.$$

The eigenvalues are $\lambda = p, q$, where $pa^2 = 1$, $qb^2 = 1$. Consider first the eigenvalue $\lambda = q$ of *larger* absolute value. Write C for the point circle with centre $F = (\alpha, \beta)$. Then

$$Q - qC = (p - q)x^2 + 2q\alpha x + 2q\beta y - (q\alpha^2 + q\beta^2 + 1).$$

To be a repeated line this expression has the form $(p - q)(x + r)^2$. Equating coefficients of x, y and the constant term gives

$$2q\alpha = 2(p - q)r, \quad 2q\beta = 0, \quad -(q\alpha^2 + q\beta^2 + 1) = (p - q)r^2.$$

Substituting the values of α, β from the first two relations into the third yields a quadratic in r, namely

$$r^2 = \frac{q}{p(q - p)}. \tag{8.4}$$

The right-hand side is *positive*, so we obtain solutions $r = \pm a/e$, giving $\alpha = \pm ae$, $\beta = 0$. That produces the foci and directrices of the lemma. We leave the reader to repeat the calculation for the eigenvalue $\lambda = p$ of *smaller* absolute value, producing

$$r^2 = \frac{p}{q(p - q)}. \tag{8.5}$$

This time the right-hand side is *negative*, so there are no solutions for r, and hence no further foci or directrices. In this way we obtain the following constructible forms for a standard real ellipse, one given by the '+' sign, and the other by the '−' sign

$$(x \pm ae)^2 + y^2 = e^2 \left(x \pm \frac{a}{e}\right)^2.$$

$\square$

Thus any standard ellipse has exactly two foci on the major axis, and two associated directrices perpendicular to that axis. Using this fact, it will follow from the listing of conics in Chapter 15 that *any* real ellipse has the same property. In practice the foci and directrices of a non-circular real ellipse can be found by mimicking the above calculation: the eigenvalue of larger absolute value gives rise to two foci and two directrices, whilst that of smaller absolute value gives rise to none.

Example 8.7 For the real ellipse $Q = 7x^2 + 2xy + 7y^2 + 10x - 10y + 7$ the characteristic equation is $\lambda^2 - 14\lambda + 48 = 0$ giving eigenvalues $\lambda = 6, 8$. For the eigenvalue $\lambda = 8$ of larger absolute value we have

$$Q - 8C = -(x - y)^2 + 2(8\alpha + 5)x + 2(8\beta - 5) - (8\alpha^2 + 8\beta^2 - 7).$$

For this to be a repeated line it has to have the form $-(x - y + r)^2$. Equating coefficients of x, y and the constant term we obtain

$$8\alpha + 5 = -r, \qquad 8\beta - 5 = r, \qquad 8\alpha^2 + 8\beta^2 - 7 = r^2.$$

Solving the first two relations for α, β in terms of r, and substituting in the third relation yields the quadratic $6r^2 - 20r + 6 = 0$ factorizing as $(r - 3)$ $(6r - 2) = 0$, so giving $r = 3, 1/3$. Substituting back we obtain the foci and directrices

$$F = (-1, 1), \quad D = x - y + 3; \qquad F' = \left(-\frac{2}{3}, \frac{2}{3}\right), \quad D' = x - y + \frac{1}{3}.$$

To determine the eccentricity we choose a focus and directrix, and compare Q with (8.2). Choosing the focus F and directrix D, and putting D into canonical form, we see that Q is a constant multiple of the quadratic function below. It follows that the eccentricity is given by $e^2 = 1/4$, so we must have $e = 1/2$

$$(x + 1)^2 + (y - 1)^2 = \frac{1}{4}\left(\frac{x - y + 3}{\sqrt{2}}\right)^2.$$

Lemma 8.5 *The standard hyperbola with moduli a, b has just two constructions, with eccentricity e defined by $a^2e^2 = a^2 + b^2$. The foci and directrices are*

$$F^+ = (ae, 0), \quad D^+ = x - \frac{a}{e}; \qquad F^- = (-ae, 2), \quad D^- = x + \frac{a}{e}.$$

Proof The analysis of the standard ellipses in Lemma 8.4 can be repeated for the standard hyperbolas

$$Q(x, y) = \frac{x^2}{a^2} - \frac{y^2}{b^2} - 1.$$

The eigenvalues are $\lambda = p, q$ defined by $pa^2 = 1, qb^2 = -1$. The calculations are identical to those for the ellipse. The eigenvalue $\lambda = q$ gives rise to the expression (8.4) for r^2 having a positive right-hand side, and producing the foci F^+, F^- and directrices D^+, D^- in the statement of the lemma. The eigenvalue $\lambda = p$ gives rise to the expression (8.5) for r^2 having a negative right-hand side, so giving no solutions.   □

Thus the standard hyperbola also has exactly two foci on one axis, and two parallel directrices perpendicular to that axis. Using this fact, it will follow from the listing of conics in Chapter 15 that *any* hyperbola has the same property. As in the case of the real ellipse, one eigenvalue gives rise to two foci and two directrices, whilst the other gives rise to none.

Example 8.8 For the hyperbola $Q = 3x^2 - 4xy + 2x + 4y - 9$ the characteristic equation is $\lambda^2 - 3\lambda - 4 = 0$ giving eigenvalues $\lambda = -1, 4$. For the eigenvalue $\lambda = -1$ we have

$$Q + C = (2x - y)^2 + 2(1 - \alpha)x + 2(2 - \beta)y + (\alpha^2 + \beta^2 - 9).$$

For this to be a repeated line it has to have the form $(2x - y + r)^2$. Equating coefficients of x, y and the constant term we obtain

$$2(1 - \alpha) = 4r, \qquad 2(2 - \beta) = -2r, \qquad \alpha^2 + \beta^2 - 9 = r^2.$$

Solving the first two relations for α, β and substituting in the third relation yields $r^2 = 1$ giving $r = \pm 1$. The value $r = 1$ gives the focus $F = (-1, 3)$ and the directrix $D = 2x - y + 1$: and the value $r = -1$ gives the focus $F' = (3, 1)$ and the directrix $D' = 2x - y - 1$. Putting the directrices into canonical form, we see that the eccentricity is given by $e^2 = 5$, so $e = \sqrt{5}$. The eigenvalue $\lambda = 4$ gives rise to a quadratic equation in r having no roots.

Exercises

8.5.1   Find the foci and the directrices of the following conics:

(i) $9x^2 - 24xy + 41y^2 - 15x - 5y = 0$,

(ii) $3x^2 - 2xy + 3y^2 - 8x + 8y + 15 = 0$,

(iii) $3x^2 + 4xy - 2x - 6y - 4 = 0$.

8.5.2   Find the foci and the directrices of the following conics:

(i) $x^2 + 8xy - 5y^2 - 2x + 6y - 6 = 0,$
(ii) $3x^2 + 2xy + 3y^2 - 6x + 14y - 101 = 0,$
(iii) $3x^2 - 10xy + 3y^2 + 8x - 24y - 8 = 0.$

8.5.3   Show that the conic $Q = 30x^2 - 12xy + 35y^2 - 24x - 16y - 16$ has one focus at the origin. Find the equation of the corresponding directrix, the eccentricity, and the coordinates of the second focus.

9

Tangents and Normals

As we stated in Chapter 4, a fundamental idea in studying a conic Q is to understand how it intersects lines. It is however not just the intersections of Q with a *single* line which are significant for its geometry, but its intersections with *pencils* of lines. That is a major theme of this text, which we introduced in Chapter 7 by studying the intersections of Q with parallel pencils of lines. In this chapter we develop the theme by studying how Q meets a general pencil of lines through a point on Q itself. That leads to a central geometric idea, the 'tangent' to Q at a point, representing the best possible first-order approximation. In Section 9.3 we introduce the companion idea of the 'normal' to Q at a point, the line through that point perpendicular to the tangent. In the next three chapters we will use the material developed so far to look at the three main conic classes of ellipses, parabolas, and hyperbolas in more detail. Each has distinctive features, which are best discussed within the context of their class.

9.1 Tangent Lines

Consider the pencil of lines through a fixed point W on a conic Q. Think of another point W' on Q, and consider the line L through W, W'. (Figure 9.1.) The idea is that as W' moves along Q into coincidence with W, so L will tend towards a limiting position, the 'tangent' line at W. To make this idea work we need to impose just one restriction on the point W, namely that it should be non-singular on Q.

Lemma 9.1 *Let Q be a conic, and let $W = (u, v)$ be a non-singular point on Q. There is a unique line T through W touching Q at W, with equation*

$$(x - u)Q_x(u, v) + (y - v)Q_y(u, v) = 0. \tag{9.1}$$

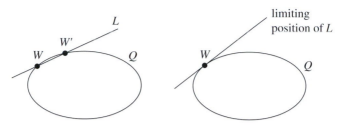

Fig. 9.1. Idea of a tangent

Proof Parametrize the line L through W with direction vector (X, Y) as $x(t) = u + tX$, $y(t) = v + tY$. Its intersections with Q are given by the intersection quadratic $pt^2 + qt + r = 0$. Since W lies on Q we have $r = Q(u, v) = 0$, so $t = 0$ is a root, corresponding to the intersection at W. The condition for L to touch Q at W (Section 4.4) is that $t = 0$ should be a repeated root, i.e. that

$$0 = q = Q_x(u, v)X + Q_y(u, v)Y.$$

The coefficients of X, Y cannot both be zero: otherwise, W is a centre of Q by Lemma 5.3, hence singular on Q, contrary to hypothesis. Thus $q = 0$ defines a *unique* direction (X, Y), namely $X = -Q_y(u, v)$, $Y = Q_x(u, v)$. To finish the proof we need only observe that the unique line through W with direction (X, Y) is that in the statement of the result. $\square$

The unique line T of Lemma 9.1 is the *tangent line* to Q at W, and (9.1) is the *tangent formula*. Of course, the concept does not apply to conics whose zero set is empty, or a single point. The mental picture is enhanced by noting that when Q is irreducible, no tangent line to Q meets Q at a further point: indeed, the intersection quadratic of Lemma 9.1 has a repeated zero t, but is not identically zero, so cannot have another zero.

Exercises

9.1.1   Show that no tangent to a conic Q can pass through a centre not lying on Q.

9.1.2   Show that the tangent at any point W on a line L is the line L itself.

9.2 Examples of Tangents

As we will see in Chapter 15, much of the detailed geometry of conics can be established via specific calculations on standard conics. That is the reason

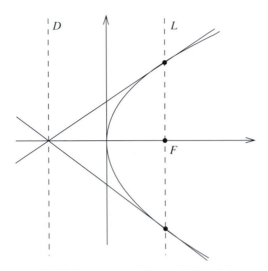

Fig. 9.2. Latus rectum of the standard parabola

for their continued prominence in the examples. Moreover, standard conics also admit particularly simple parametrizations, frequently simplifying calculations. The following examples illustrate these points, adding to our store of knowledge.

Example 9.1 Consider the standard parabola $Q(x, y) = y^2 - 4ax$ with $a > 0$, and a point (u, v) on Q, so satisfying $v^2 = 4au$. Then $Q_x = -4a$, $Q_y = 2y$, and the tangent formula shows that the tangent at (u, v) has equation $vy = 2a(u + x)$.

A typical illustration is provided by the geometry of the 'latus rectum'. Recall that a focus F of a constructible conic Q always lies on an axis. The chord L of Q through F perpendicular to that axis is known as a *latus rectum* of Q. Thus a parabola has a unique latus rectum, whilst the real ellipse and hyperbola have two. Such chords have interesting geometric properties.

Example 9.2 For the standard parabola with modulus a the focus is $F = (a, 0)$ so the latus rectum L is the chord $x = a$. We claim that the tangents at the ends of the chord pass through the point where the axis $y = 0$ meets the directrix D defined by $x = -a$. (Figure 9.2.) Indeed, the ends of the chord are the points $(a, \pm 2a)$ with tangents the lines $y = \pm(x + a)$. Thus the tangents have slopes ± 1, and pass through the required point $(-a, 0)$.

Example 9.3 Another illustration arises when we consider the standard parabola parametrized as $x(t) = at^2$, $y(t) = 2at$. In that case the tangent at the point with parameter t has equation

$$x - ty + at^2 = 0.$$

From this formula one can immediately read off the fact that points on the parabola with different parameters have tangents with different directions. Thus the standard parabola has no parallel tangents. By contrast, the tangents at the ends of a diameter of an ellipse or hyperbola are automatically parallel. Here is another application of the formula, establishing a useful geometric property of the parabola, namely that a chord passes through the focus if and only if the tangents at the ends are perpendicular.

Example 9.4 Consider the standard parabola $y^2 = 4ax$ with modulus a parametrized as $x(t) = at^2$, $y(t) = 2at$. By Example 9.3 the tangents at the points with parameters t_1, t_2 are the following lines with respective slopes $1/t_1$, $1/t_2$

$$\begin{cases} x - t_1 y + at_1^2 = 0 \\ x - t_2 y + at_2^2 = 0. \end{cases}$$

By Exercise 4.2.3 the chord through the points with parameters t_1, t_2 has equation

$$(t_1 + t_2)(y - 2at_1) = 2(x - at_1^2).$$

It remains to observe that the condition for the chord to pass through the focus $(a, 0)$ is that $t_1 t_2 = -1$: and by Example 6.5 that is the condition for the tangents to be perpendicular.

A potentially interesting question is how many tangents to a given conic pass through a fixed point in the plane. For the standard parabola that produces another typical application of the formula for the tangent to the parametrized curve.

Example 9.5 For the standard parabola $Q(x, y) = y^2 - 4ax$ with modulus a, parametrized as $x(t) = at^2$, $y(t) = 2at$, we saw that the tangent line at the point (x, y) is $x - ty + at^2 = 0$. The tangent passes through a fixed point $P = (u, v)$ if and only if t satisfies the quadratic equation $u - tv + at^2 = 0$. The number of roots is determined by its discriminant $v^2 - 4ax$: there are two roots when $v^2 > 4ax$ (P is outside the parabola), just one when $v^2 = 4ax$ (P lies on the parabola), and none when $v^2 < 4ax$ (P lies inside the parabola). For instance, when $u = -a$, $v = 0$ the point P is the intersection of the axis with the directrix, so outside the parabola. The quadratic then has roots $t = \pm 1$ producing the ends $(a, \pm 2a)$ of the latus rectum, agreeing with the result of Example 9.2.

Example 9.6 Consider conics $Q(x, y) = \alpha x^2 + \beta y^2 + \gamma$ with α, β, γ non-zero, so having no singular points. We will show that the tangent at a point $w = (u, v)$ on Q has equation

$$\alpha u x + \beta v y + \gamma = 0.$$

This is particularly easy to remember, since one has only to replace x^2, y^2 in Q by ux, vy respectively. The result is easily verified. Since w lies on Q we have $\alpha u^2 + \beta v^2 + \gamma = 0$: also, $Q_x(x, y) = 2\alpha x$, $Q_y(x, y) = 2\beta y$ so by the tangent formula the tangent is

$$0 = (x - u)(2\alpha u) + (y - v)(2\beta v) = 2(\alpha u x + \beta v y + \gamma).$$

Example 9.7 A particular case of the previous example is provided by the standard ellipses and hyperbolas of Examples 4.2, 4.3 with equations

$$\frac{x^2}{a^2} + \frac{y^2}{b^2} = 1, \qquad \frac{x^2}{a^2} - \frac{y^2}{b^2} = 1.$$

In those cases the tangents at a point (u, v) have respective equations

$$\frac{ux}{a^2} + \frac{vy}{b^2} = 1, \qquad \frac{ux}{a^2} - \frac{vy}{b^2} = 1.$$

Example 9.8 For the standard ellipse with moduli a, b, parametrized as $x(t) = a \cos t$, $y(t) = b \sin t$, the tangent at the point with parameter t has equation

$$bx \cos t + ay \sin t = ab.$$

Likewise, for the standard hyperbola with moduli a, b, parametrized as $x(t) = a \cosh t$, $y(t) = b \sinh t$, the tangent at the point with parameter t has equation

$$bx \cosh t - ay \sinh t = ab.$$

The concept of tangency can extended in a natural way. Two conics Q, Q' are *tangent* at a common non-singular point P when the tangents to Q, Q' at P coincide.

Example 9.9 The major and minor auxiliary circles of the standard ellipse E with moduli a, b were defined to be the concentric circles of radii a, b. (Figure 5.3.) The major auxiliary circle meets E only at the vertices $(\pm a, 0)$ on the major axis, and shares common tangents $x = \pm a$. Likewise, the minor auxiliary circle meets E only at the vertices $(0, \pm b)$, and shares common tangents $y = \pm b$. Thus the auxiliary circles are tangent to E at the vertices.

Exercises

9.2.1   In each of the following cases find the tangent to the conic Q at the point P:

    (i) $Q = x^2 - 6y^2 - 3$, $P = (3, 1)$,
    (ii) $Q = 2x^2 + 3y^2 - 11$, $P = (2, 1)$,
    (iii) $Q = (x - y - 1)^2 - 4y$, $P = (1, 0)$.

9.2.2   Find the common parallel tangents of the circles of radius 1 having centres $(1, 0)$, $(0, 2)$.

9.2.3   Let L, M be lines in different directions, intersecting at a single point. Show that the tangent to the line-pair L, M at any other point on L (respectively M) is L (respectively M). This exercise shows that *a tangent line to a conic can have an asymptotic direction.*

9.2.4   Show that the tangents to a conic Q at the ends of a diameter are parallel.

9.2.5   Let a, b be positive constants. A circle of radius b touches the parabola $y^2 = 4ax$ at the vertex. Find the coordinates of the other intersections of the two conics.

9.2.6   Show that there is a fixed line L such that the tangents at points with parameter t on the ellipse $x(t) = a \cos t$, $y(t) = b \sin t$ and the hyperbola $x(t) = a \sec t$, $y(t) = b \tan t$ intersect on L.

9.2.7   Show that the hyperbola $xy = c^2$ intersects the standard ellipse with moduli a, b in four, two, or no points according as $ab > 2c^2$, $ab = 2c^2$, or $ab < 2c^2$. In the second case, show that the conics are tangent at the intersections.

9.2.8   Let E be a standard ellipse with eccentricity e. Show that the tangents at the ends of a latus rectum have slopes $\pm e$, pass through the intersection of the major axis with a directrix, and pass through an intersection of the minor axis with the major auxiliary circle.

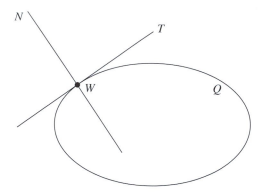

Fig. 9.3. A normal line to an ellipse

9.3 Normal Lines

Let Q be a conic, and let $W = (u, v)$ be a non-singular point on Q. As we saw in the previous section, there is a unique tangent line T at W, meeting Q at no other point. The line through W perpendicular to T is the *normal line* N to Q at W, and we refer to W as the *foot* of the normal. By Lemma 9.1 the normal line to a conic Q at a non-singular point $W = (u, v)$ has the equation

$$(x - u)Q_y(u, v) - (y - v)Q_x(u, v) = 0. \qquad (9.2)$$

Our first example illustrates a very basic fact in the geometry of the circle, namely that all the normals to a real circle pass through its centre.

Example 9.10 Consider the real circle of radius r with centre (α, β) given by

$$C(x, y) = (x - \alpha)^2 + (y - \beta)^2 - r^2.$$

Using (9.2) we see that the normal at any point (u, v) on the circle is the line

$$(x - u)(v - \beta) - (y - v)(u - \alpha) = 0.$$

Clearly, this relation is satisfied by $x = \alpha$, $y = \beta$ so the normal line passes through the centre.

Example 9.11 Consider the parametrization $x(t) = a \cos t$, $y(t) = b \sin t$ of the standard ellipse with moduli a, b. The normal at the point with parameter t has equation

$$ax \sin t - by \cos t = (a^2 - b^2) \sin t \cos t.$$

Thus the normal at the point with parameter t passes through the centre if and only if $\sin t = 0$ or $\cos t = 0$: these relations hold at exactly four points on the ellipse, namely the vertices $(\pm a, 0)$, $(0, \pm b)$. More generally, it can be shown that there are at most four points on an ellipse at which the normal passes through a given point.

Example 9.12 Let Q be the standard parabola $Q(x, y) = y^2 - 4ax$ with $a > 0$. Using (9.2) we see that the normal at a point (u, v) on Q is

$$vx + 2ay = 2v(u + 2a).$$

In particular, when Q is parametrized as $u(t) = at^2$, $v(t) = 2at$, the normal at t has equation

$$tx + y = at^3 + 2at.$$

In the preceding section we asked for the number of *tangents* to the standard parabola through a given point (u, v), and found an answer that accorded well with our mental picture for a parabola. It is much more interesting to ask for the number of *normals* through (u, v).

Example 9.13 In the previous example we saw that the normal to the standard parabola Q, parametrized as $x(t) = at^2$, $y(t) = 2at$, is $tx + y = at^3 + 2at$. Thus the values of t for which the normal passes through (u, v) are given by the cubic equation

$$at^3 + (2a - u)t - v = 0. \tag{9.3}$$

The number of roots depends on the choice of (u, v). In principle, a cubic has either three distinct real roots, or two distinct roots (one repeated), or no real roots. The repeated root case is transitional between the other two, and one expects it to hold on a *curve* in the (u, v)-plane. We will determine this curve explicitly. To this end, note that the displayed cubic has no term in t^2, so by polynomial algebra the sum of its roots is zero. It follows that when we have a repeated root t, the other root must be $-2t$. The normals at these points are

$$tx + y = at^3 + 2at, \qquad -2tx + y = -8at^3 - 4at.$$

We leave the reader to check that the normals intersect at the point $u = a(3t^2 + 2)$, $v = -2at^3$. Eliminating t from these relations we find that points on the transitional curve satisfy the equation of a cubic curve

$$4(u - 2a)^2 = 27av^2. \tag{9.4}$$

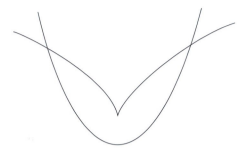

Fig. 9.4. Evolute of a parabola

Figure 9.4 shows the cubic curve (9.4) superimposed on an illustration of the standard parabola. It is called the *semicubical parabola*. The cubic is symmetric about the u-axis, cuts the parabola twice when $u = 8a$, and exhibits a 'cusp' when $u = 2a$. Three distinct normals pass through any point inside the cubic, whilst none pass through points outside: points on the cubic lie on two normals, with the sole exception of the 'cusp', which lies on just one. The concept of a cusp does not arise for conics, but is a significant feature in the algebraic geometry of general plane curves, discussed in *EGAC*. The semicubical parabola appears in the differential geometry of plane curves, as a special case of the 'evolute' construction, explained in *EGDC*.

Exercises

9.3.1   In each of the following cases find the normal to the conic Q at the point P:

$$
\begin{array}{lll}
\text{(i)} & Q = x^2 + y^2 - 2x + 4y - 20, & P = (-2, -2), \\
\text{(ii)} & Q = 2x^2 + 2y^2 - 4x + 9y, & P = (0, 0), \\
\text{(iii)} & Q = (x - y - 1)^2 - 4y & P = (1, 0).
\end{array}
$$

9.3.2   Show that if all the normals to a conic Q pass through a fixed point P, then Q is a circle, and P is its centre.

9.3.3   Consider the normals to the standard parametrized parabola $x(t) = at^2$, $y(t) = 2at$ passing through a point $(u, 0)$ on the axis. Verify that for $u > 2a$ there are three distinct normals, that for $u = 2a$ there is just one, and that for $u < 2a$ there are none.

9.3.4   Find the coordinates of the point where the normal at the point with parameter t on the standard parabola $x(t) = at^2$, $y(t) = 2at$ meets the parabola again.

9.3.5   Consider a parallel pencil of lines $x = my + c$, with m fixed. Show that the parameters t_1, t_2 of the points on the standard parametrized parabola $x(t) = at^2$, $y(t) = 2at$, where it meets a line $x = my + c$, are related by $t_1 + t_2 = 2m$. Use this fact to show that the normals to the parabola, at its intersections with any line in the pencil, intersect on the normal at a point with fixed parameter t_3.

9.3.6   Let Q be a standard parabola, and let P be a point through which three normals to Q can be drawn. Show that the feet of the normals lie on a circle through the vertex.

9.3.7   Show that the normal at any point on a standard ellipse Q meets the major axis at a point between the two foci.

9.3.8   Let H be a standard hyperbola, and let Y be the half-plane $y > 0$. Show that for any point P in Y there are exactly two points in Y on H at which the normals pass through P.

9.3.9   A *focal chord* of a non-degenerate conic Q is a chord passing through a focus. Through any point P on a standard ellipse Q there are two focal chords, meeting the ellipse again at points P', P''. Show that the tangents to Q at P', P'' intersect on the normal at P.

9.3.10   A focal chord intersects a standard ellipse E at the points P, Q. The tangents at P, Q intersect at R, and the normals at S. Show that RS passes through the other focus of E.

10

The Parabola

The parabola occupies a transitional position between the ellipse and the hyperbola, with its own distinctive geometry. In several respects it is easier to deal with than ellipses and hyperbolas, and for that reason we discuss it first. For instance, it has but a single axis of symmetry, whereas ellipses and hyperbolas have two. On that basis we develop a simple, efficient technique for determining the main geometric features in Section 10.2. Finally, we use these techniques to establish a unique feature of the parabola, that it admits a parametrization by quadratic functions.

10.1 The Axis of a Parabola

For irreducible conics (circles excepted) the direction quadratic gives two perpendicular directions, which *may* give rise to an axis. For conics with a unique centre (in particular, ellipses and hyperbolas) we can say more. Lemma 10.1 guarantees that both directions give rise to axes. However, for parabolas we need to think again. The illustration of the standard parabola in Figure 8.1 suggests that it has only *one* axis, intersecting the parabola at a unique vertex. The next result verifies that is the case for any parabola. As usual we consider a general conic

$$Q(x, y) = ax^2 + 2hxy + by^2 + 2gx + 2fy + c. \qquad (\star)$$

Lemma 10.1 *Any parabola Q has a unique axis, of the form $ax + hy + k$ or $hx + by + k$, for some constant k. Moreover, the axis intersects Q in a unique vertex.*

Proof Since Q is a parabola we have $\delta = ab - h^2 = 0$. Thus the direction quadratic $aX^2 + 2hXY + bY^2 = 0$ is a perfect square, with just one root up to

98

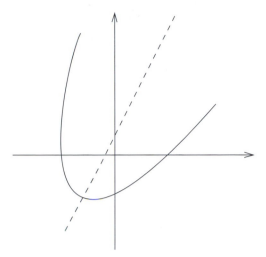

Fig. 10.1. The conic Q of Example 10.1

constant multiples. We can determine a root explicitly. At least one of the vectors $(-h, a)$, $(-b, h)$ is non-zero, and they are linearly dependent. Moreover, they satisfy the direction quadratic, i.e. it is satisfied when $X = -h$, $Y = a$, or $X = -b$, $Y = h$. By (7.1) the midpoint locus associated to the perpendicular directions (a, h), (h, b) are defined by the relations

$$\begin{cases} (a^2 + h^2)x + h(a + b)y + (ga + fh) = 0 \\ h(a + b)x + (h^2 + b^2)y + (gh + fb) = 0. \end{cases}$$

We leave the reader to check that these equations agree with those in the statement of the lemma, up to constant multiples. $\square$

Example 10.1 The conic $Q(x, y) = 4x^2 - 4xy + y^2 - 10y - 19 = 0$ is a parabola, with quadratic terms $(2x - y)^2$. The axis direction is $(2, -1)$, and (7.2) shows that the axis is $2x - y + 1 = 0$. (Figure 10.1.)

10.2 Practical Procedures

The starting point for analyzing the geometry of parabolas is the following result, that parabolas can be written in a particularly useful form.

Lemma 10.2 *Any parabola Q can be written $Q = L^2 + L'$, where L, L' are lines in different directions.*

Proof By definition, a parabola has invariants $\delta = 0$, $\Delta \neq 0$. Since $\delta = 0$ the quadratic terms form a perfect square, so $Q = L^2 + L'$, with $L = \alpha x + \beta y + \gamma$, $L' = \alpha' x + \beta' y + \gamma'$ lines. By Lemma 5.3, centres for Q are given by the relations

$$0 = Q_x = 2\alpha L + \alpha', \qquad 0 = Q_y = 2\beta L + \beta'.$$

If L, L' have the same direction the vectors (α, β), (α', β') are linearly dependent, and there is a line of centres. But then Theorem 5.4 gives $\Delta = 0$, a contradiction establishing the result. □

Example 10.2 Parabolas do not have asymptotes. By Lemma 10.2 any parabola Q can be written in the form $Q = L^2 + L'$, where L, L' are lines in different directions. An asymptote must be parallel to the axis, hence parallel to L: thus it would have the form $L + k$, for some constant k. Since L, L' intersect, the parallel lines $L + k$, $L' + k^2$ likewise intersect. However, the intersection satisfies the relations $L + k = 0$, $Q = L^2 + L' = 0$, so is a point where $L + k$ meets the parabola. It follows that no line $L + k$ parallel to L can be an asymptote.

Now consider a general parabola Q, given by ($\star$). Since $\delta = 0$, at least one of a, b is non-zero. We can assume $a \neq 0$: then $aQ = L^2 + L'$ where

$$L(x, y) = ax + hy, \qquad L'(x, y) = a(2gx + 2fy + c).$$

By Lemma 10.1, the axis of Q has the form $M = L + k$, for some constant k. One line of working suffices to check that we can then write Q in the form $C = M^2 + M'$, where M, M' are given as follows. The question then is how to choose k to ensure that M is the axis

$$M = L + k, \qquad M' = L' - 2kL - k^2. \tag{10.1}$$

Lemma 10.3 *The line M is an axis for $Q = M^2 + M'$ if and only if M, M' are perpendicular. In particular, the axis M intersects Q in a unique vertex.*

Proof The relation (7.2) shows that $M = L + k$ is an axis if and only if k has the value displayed below: and the reader will readily check that this is the unique value of k for which the lines M, M' are perpendicular

$$k = \frac{a(ag + hf)}{a^2 + h^2}.$$

Clearly, an intersection of M, Q is an intersection of M, M', and conversely. (That holds for *any* choice of k.) But M, M' intersect when $L = -k$, $L' = $

$-k^2$: and since L, L' have different directions these relations have a unique solution. In particular, that is the case when M is the axis, when the intersection is the unique vertex. □

This result provides a simple technique for analysing parabolas. In a nutshell it is this. Given a parabola $Q = L^2 + L'$, find the unique value of k for which the lines M, M' defined in (10.1) are perpendicular: then M is the axis, and the intersection of M, M' is the vertex. Starting from Q, we have to do little more than solve a single linear equation in k.

Example 10.3 The parabola $Q(x, y) = 4x^2 - 4xy + y^2 - 10y - 19$ can be written $Q = L^2 + L'$, where $L(x, y) = 2x - y$, $L' = -10y - 19$. With the above notation we have

$$M = 2x - y + k, \qquad M' = -4kx + 2(k - 5)y - (k^2 + 19).$$

By calculation, M, M' are perpendicular if and only if $k = 1$. Thus $M = 2x - y + 1$, $M' = -4(x + 2y - 5)$. The vertex is the point where $M = 0$, $M' = 0$, namely $x = 3/5$, $y = 11/5$.

Exercises

10.2.1   In each of the following cases show that the given conic is a parabola, and determine its axis and vertex:

    (i) $4x^2 - 4xy + y^2 - 10y - 19 = 0$,
    (ii) $(x - y - 1)^2 - 4y = 0$,
    (iii) $x^2 - 2xy + y^2 + 2x + 3y + 2 = 0$.

10.2.2   In each of the following cases show that the given conic is a parabola, and determine its axis and vertex:

    (i) $x^2 - 4xy + 4y^2 + 10x - 8y + 13 = 0$,
    (ii) $x^2 - 4x + 4y^2 - 5y - 1 = 0$,
    (iii) $4x^2 - 4xy + y^2 - 8\sqrt{5}x - 16\sqrt{5}y = 0$.

10.2.3   In each of the following cases show that the given conic is a parabola, and determine its axis and vertex:

    (i) $y = ax^2 + 2bx + c$,
    (ii) $x^2 - 2xy + y^2 - 6\sqrt{2}x - 2\sqrt{2}y - 6 = 0$,
    (iii) $4x^2 - 4xy + y^2 + 8\sqrt{5}x + 6\sqrt{5}y - 15 = 0$.

10.3 Parametrizing Parabolas

In Example 4.4 we parametrized the standard parabola by considering its intersection with the pencil of lines parallel to its axis, so each line meets the parabola just once. The principle applies to *any* parabola Q. As in Section 10.1 we write Q in the form $Q = M^2 + M'$ with M, M' perpendicular lines. The pencil of lines parallel to the axis M comprises the lines $M + 2t$ with t any scalar. Then Q meets a line in the pencil when $M = -2t$, $Q = 0$, or equivalenty $M = -2t$, $M' = -4t^2$. Since M, M' are perpendicular, these linear equations in x, y have a unique solution, providing a parametrization of the form

$$x(t) = x_0 + x_1 t + x_2 t^2, \qquad y(t) = y_0 + y_1 t + y_2 t^2. \qquad (10.2)$$

Example 10.4 The parabola of Example 10.3 was written $Q = M^2 + M'$ with M, M' perpendicular lines, namely

$$M(x, y) = 2x - y + 1, \qquad M'(x, y) = -4(x + 2y - 5).$$

We obtain a parametrization by solving the linear equations $M = -2t$, $M' = -4t^2$. Written out in full

$$2x - y + 1 = -2t, \qquad -4(x + 2y - 5) = -4t^2.$$

We leave the reader to verify that the resulting parametrization is

$$x(t) = \frac{3 - 4t + t^2}{5}, \qquad y(t) = \frac{11 + 2t + 2t^2}{5}.$$

Here is an interesting illustration of the value of parametrization, namely the remarkable reflective property of a parabola.

Example 10.5 Consider the standard parabola $y^2 = 4ax$ with focus $F = (a, 0)$. We claim that for any point P on the parabola the perpendicular bisectors of the line L through F, P and the 'horizontal' line M through P are the tangent and normal lines at P. It suffices to show that the tangent is parallel to a bisector. We can parametrize the parabola as in (4.4) so $P = (at^2, 2at)$. Then the lines are

$$L = 2tx + (1 - t^2)y - 2at, \qquad M = y - 2at.$$

Deleting the constant terms and multiplying out, we see that the parallel lines through the origin have joint equation $2txy + (1 - t^2)y^2$. Thus the joint

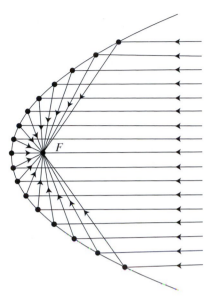

Fig. 10.2. Reflective property for a parabola

equation of the perpendicular bisectors is

$$tx^2 + (1 - t^2)xy - ty^2 = (x - ty)(tx + y).$$

It remains to observe that by Example 9.3 the tangent line at P is the line $x - ty + at^2 = 0$, which is parallel to the first factor.

That establishes the reflective property of the parabola, that for any incident ray through the focus, the reflected ray is parallel to the axis. Thus the pencil of incident rays through the focus gives rise to a pencil of reflected rays parallel to the axis. That fact is of enormous practical importance for line-of-sight radio communications at ultra high frequencies. Parabolic dishes, obtained by rotating a parabola about its axis, are now a ubiquitous part of the landscape: they are mounted on houses to receive satellite TV, on masts to provide permanent microwave links for emergency and public services, whilst giant versions are erected outside urban areas for radio astronomy research and early warning missile systems. In all these cases a wavefront can be propagated or received via an antenna at the focus. In optics the same property is used in torches and searchlights to produce a parallel beam of high intensity light.

Exercises

10.3.1 Show that the parametrization $x = at^2$, $y = 2at$ of a standard parabola $y^2 = 4ax$ with $a > 0$ can also be derived by considering the pencil of non-horizontal lines $2x = ty$ through $p = (0, 0)$.

10.3.2 The line $\alpha x + \beta y + a\gamma = 0$ meets the standard parabola $y^2 = 4ax$ with focus F at the points P, Q. And the lines through F, P and F, Q meet the parabola at R, S. Show that the line joining R, S is $\gamma x - \beta y + a\alpha = 0$.

10.3.3 Let C be a standard parabola with focus F and directrix D. For distinct points P, Q on C let R be the point where the line PQ meets the axis, and let S be the point where the tangents intersect. Show that the distance from R to F coincides with the distance from S to D.

10.3.4 Show that if $a > b > 0$ and $c > 2(a - b)$ then the parabolas $y^2 = 4a(x + c)$ and $y^2 = 4bx$ have two common normal lines.

10.3.5 Show that the normals at the ends of a focal chord of a parabola $y^2 = 4ax$ intersect on the parabola $y^2 = a(x - 3a)$.

10.3.6 In each of the following cases find a parametrization of the given parabola of the form (10.2):

(i) $y = ax^2 + 2bx + c$,
(ii) $x^2 + 2xy + y^2 - 6x - 2y + 4 = 0$,
(iii) $(x - y - 1)^2 = 4y$.

11

The Ellipse

The geometry of the ellipse differs substantially from that of the parabola, since it has two axes of symmetry (whereas the parabola has just one) and is a central conic (whereas the parabola is not). Our first result is that all the lines passing through the centre meet the ellipse in two distinct points, distinguishing the ellipse visually from the hyperbola, and establishing the existence of exactly four vertices.

In Section 11.2 we take up the question of parametrization. Unlike parabolas, it is not possible to parametrize general ellipses by quadratic functions of a single variable. However ellipses can be parametrized in terms of rational functions, quotients of polynomial functions. Such rational parametrizations have interesting applications to other areas of mathematics. By way of illustration we indicate how to solve a problem of ancient Greek mathematics, that of listing right-angled triangles with integer sides.

The remainder of the chapter is devoted to focal properties of ellipses, in particular the interesting metric property that the sum of the distances from any point on the ellipse to the two foci is constant. The final section establishes a reflective property for ellipses, analogous to that for parabolas.

11.1 Axes and Vertices

Perhaps one of the most obvious properties of the circle is that every line through the centre cuts the circle twice. The ellipse should be thought of as a natural generalization of the circle, so one expects it to have the same property. Indeed that is the case.

Lemma 11.1 *Every line L through the centre of a real ellipse Q meets Q in exactly two distinct points.*

Proof Let (u, v) be the centre. As in Section 1.3 we can parametrize L as $x(t) = u + tX$, $y(t) = v + tY$. Then intersections with Q are given by the roots of the quadratic equation (4.2)

$$\phi(t) = pt^2 + qt + r = 0.$$

The coefficients are given by the formulas (4.3). Since (u, v) is a centre $Q_x(u, v) = 0$, $Q_y(u, v) = 0$, and the formulas show that $q = 0$. Thus intersections are given by the roots of a quadratic $\phi(t) = pt^2 + r = 0$. The constant term $r = Q(u, v)$ is non-zero, since the centre does not lie on Q. Moreover the sign of p is independent of the direction (X, Y), since the assumption $\delta > 0$ means that $p(X, Y)$ does not vanish. Thus $\phi(t) = 0$ *either* has no roots for any choice of direction, *or* it has two distinct roots for any choice. We must have the latter possibility: by assumption there is a point in the zero set of Q, and the line L joining it to the centre meets Q.   □

The results of Chapter 7 show that any conic Q with non-zero delta invariant (circles excepted) has two distinct non-zero eigenvalues, giving rise to perpendicular axes through the centre. In particular, that is the case for the ellipse, each axis meeting the ellipse at two vertices equidistant from the centre. The distance between the two vertices on an axis is the *length* of that axis, and the distance from the centre to a vertex is its *semilength*. The axis of greater (resp. smaller) length is the *major* (resp. *minor*) axis, and corresponds to the eigenvalue of smaller (resp. larger) absolute value. (Section 7.2.) The two circles concentric with an ellipse passing through the vertices are the *auxiliary circles* associated to Q: the *minor* (resp. *major*) auxiliary circle is that of smaller (resp. larger) radius.

It follows from the above that *any ellipse has four distinct vertices*. It turns out that 'vertices' are important concepts in their own right in the differential geometry of plane curves, with conics providing the first interesting examples. Thus, even in the limited arena of conics, we see concepts of general differential geometry making their first tentative appearance.

Exercises

11.1.1   Let C, C' be real circles, centre the origin of radii r, r' with $r < r'$. Any half-line emanating from the origin meets C, C' at unique points Q, Q'. Let P be the intersection of the vertical line through Q with the horizontal line through Q'. Show that the locus of P is the zero set of a standard ellipse.

11.1.2   Let E be a standard ellipse, and let A^+, A^- be the ends of the major axis. For any point P on E (distinct from A^+, A^-) let P^+, P^- be the intersections of the tangent at P with the tangents at A^+, A^-. Show that for any two distinct points P, Q on E the lines P^+Q^-, P^-Q^+ intersect on the major axis.

11.1.3   Let K be a point on the major axis of a standard ellipse E. A line through K meets the ellipse at points P, Q. Show that the tangents at P, Q intersect the line through K parallel to the minor axis at points equidistant from K.

11.1.4   Let E be an ellipse, let P be a point on E, and let Q, R be the projections of P on to the axes. Show that there is an ellipse E' concentric with E such that the lines QR are normals to E'.

11.2 Rational Parametrization

It is tempting to wonder whether the regular parametrization of the parabola in Section 10.3 by quadratic functions of the following form can be extended to the ellipse

$$x(t) = x_0 + x_1 t + x_2 t^2, \qquad y(t) = y_0 + y_1 t + y_2 t^2. \qquad (11.1)$$

Example 11.1 That certainly does not work for the unit circle defined below

$$C(x, y) = x^2 + y^2 - 1.$$

Suppose indeed that the functions $x(t)$, $y(t)$ of (11.1) parametrize C. Substitute $x(t)$, $y(t)$ for x, y in $C(x, y) = 0$. Equating coefficients of t^4 gives $x_2^2 + y_2^2 = 0$, so $x_2 = y_2 = 0$: then, equating coefficients of t^2 we obtain $x_1^2 + y_1^2 = 0$ and hence $x_1 = y_1 = 0$. It follows that any parametrization (11.1) would be constant.

At this point a little thought goes a long way. The virtue of parametrizing a parabola by the pencil of lines parallel to its axis is that the lines meet the parabola *just once*. Actually, the mental picture is that all the lines in the pencil meet the conic *twice*, once in the Euclidean plane and once at the 'point at infinity'. That motivates the following geometric approach, applicable to more general conics Q. Consider the intersections of Q with the pencil of lines L through a fixed point P on Q. The lines L are parametrized by their slope t. In principle, L meets Q in two points, namely the point P itself, and some other point whose coordinates depend on t. Here is an example.

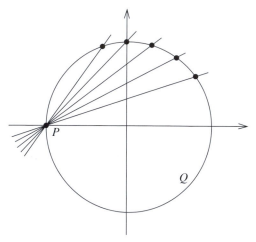

Fig. 11.1. Rational parametrization of the circle

Example 11.2 Consider the case when Q is the circle below, and P is chosen to be the point $P = (-a, 0)$, where $a > 0$

$$Q(x, y) = x^2 + y^2 - a^2.$$

Any 'non-vertical' line through P has the form $y = t(x + a)$ for some scalar t and meets the circle at P, and at some other point whose coordinates depend on t. (Figure 11.1.) To find the second point, substitute in Q to obtain a quadratic in x

$$(1 + t^2)x^2 + 2at^2x + a^2(t^2 - 1) = 0.$$

Observe that $x = -a$ must be a root of this quadratic, so $(x + a)$ is a factor. By inspection, the other factor is $(1 + t^2)x + a(t^2 - 1)$, so the coordinates of the second point of intersection are

$$x(t) = a \left(\frac{1 - t^2}{1 + t^2} \right), \qquad y(t) = t(x(t) + a) = \frac{2at}{1 + t^2}.$$

Two observations are worth making. The first is that P is the *only* point on the circle that does not correspond to any value of t. And the second is that the result can be deduced from the standard parametrization $x(t) = a \cos \theta$, $y(t) = a \sin \theta$ using the half-angle formulas displayed below. Attractive though this alternative derivation may be, it has the drawback of hiding the underlying geometric idea

$$t = \tan \frac{\theta}{2}, \qquad \cos \theta = \frac{1 - t^2}{1 + t^2}, \qquad \sin \theta = \frac{2t}{1 + t^2}.$$

Although the functions $x(t)$, $y(t)$ in Example 11.2 are not of the form (11.1) they are *quotients* of such functions. Thus they are examples of *rational* functions of the form $u(t)/v(t)$, where $u(t)$, $v(t)$ are polynomials in t. Rational parametrizations of conics have applications which are useful, interesting and quite unexpected. Here is an application to a classical problem in number theory.

Example 11.3 A problem of ancient Greek mathematics was that of finding all right-angled triangles with integer sides, i.e. all positive integers X, Y, Z with $X^2 + Y^2 = Z^2$. The reader is probably familiar with the $3, 4, 5$ triangle of school mathematics. It is interesting to ask whether there are other examples, indeed how one might generate all possible examples. The key observation is that for such a triangle the point (x, y) on the circle $x^2 + y^2 = 1$ defined by $x = X/Z$, $y = Y/Z$ is a *rational* point (meaning that x, y are both rational numbers) in the positive quadrant; conversely, any such point (x, y) with $x = X/Z$, $y = Y/Z$ where X, Y, Z are positive integers gives rise to a triangle with the required properties. Thus the problem is that of determining rational points on the circle. In principle we know how to generate all points on the circle, via the parametrization of Example 11.2

$$x(t) = \frac{1 - t^2}{1 + t^2}, \qquad y(t) = \frac{2t}{1 + t^2}.$$

That suggests we ask for those values of t which ensure that $x(t)$, $y(t)$ are rational. Clearly, if t is rational then $x(t)$, $y(t)$ are rational: conversely, if $x(t)$, $y(t)$ are rational the first displayed formula shows that t^2 is rational, and the second shows that t is rational. We can therefore assume t is rational, say $t = u/v$ with u, v coprime integers: and for (x, y) to lie in the positive quadrant we need $0 < t < 1$, i.e. $0 < u < v$. As u, v vary so we obtain *infinitely many* solutions to our problem

$$X = v^2 - u^2, \qquad Y = 2uv, \qquad Z = v^2 + u^2.$$

The first six entries in the table below are obtained by taking $u < v \leq 5$, the first being the $3, 4, 5$ triangle of school geometry. The last two entries are of historical interest: they appear on a Babylonian clay tablet dating to around 2000BC.

u	v	X	Y	Z
1	2	3	4	5
2	3	5	12	13
1	4	15	8	17
3	4	7	24	25
2	5	21	20	29
4	5	9	40	41
4	9	65	72	97
5	12	119	120	169

Exercises

11.2.1   Let a be a non-zero constant. By considering the pencil of lines through the point $P = (0, 0)$ show the circle of radius $|a|$ centred at the point $(a, 0)$ has the rational parametrization

$$x(t) = \frac{2a}{1 + t^2}, \qquad y(t) = \frac{2at}{1 + t^2}.$$

11.2.2   Let a be a non-zero constant. By considering the pencil of lines through the point $P = (2a, 0)$ show that the circle of radius $|a|$ centred at the point $(a, 0)$ has the rational parametrization

$$x(t) = \frac{2at^2}{1 + t^2}, \qquad y(t) = \frac{2at}{1 + t^2}.$$

11.2.3   Starting from the parametrization $x(t) = a \cos t$, $y(t) = b \sin t$ of the standard ellipse with moduli a, b use the half-angle formulas to derive a rational parametrization.

11.2.4   By considering the pencil of lines through the point $P = (-a, 0)$, show that the standard ellipse with moduli a, b has the rational parametrization

$$x(t) = a \left(\frac{b^2 - a^2 t^2}{b^2 + a^2 t^2} \right), \qquad y(t) = \frac{2ab^2 t}{b^2 + a^2 t^2}.$$

11.3 Focal Properties

In Chapter 8 we saw that general ellipses have exactly two constructions, each obtained from the other by central reflexion in the centre. There are two foci on the major axis: the corresponding directrices are perpendicular to the major axis, and their parallel bisector is the minor axis.

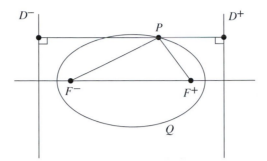

Fig. 11.2. Metric property of an ellipse

Example 11.4 The foci of an ellipse can lie inside, on, or outside the minor auxiliary circle. The transitional case, when the foci lie on the circle, is *Fagnano's ellipse*. It has a simple interpretation in terms of the eccentricity e. Consider the standard ellipse with moduli a, b. For Fagnano's ellipse the foci $F^+ = (ae, 0)$, $F^- = (-ae, 0)$ must be distance b from the centre, so $ae = b$. However, the eccentricity e is defined by the relation $a^2 e^2 = a^2 - b^2$. It follows from these relations that $e^2 = 1/2$, so that $e \sim 0.707$.

The next example describes an interesting metric property of ellipses which has no counterpart for the parabola.

Example 11.5 Let Q be the standard ellipse with moduli a, b satisfying $0 < b < a$, and eccentricity e. The foci are denoted F^+, F^- and the corresponding directrices D^+, D^-. We use the fact that the distance between the directrices is $2a/e$. Now let P be any point on Q. Since $PF^+ = ePD^+$, $PF^- = ePD^-$ we have

$$PF^+ + PF^- = e\{PD^+ + PD^-\} = e\{2a/e\} = 2a. \qquad (11.2)$$

Thus the sum of the distances from the foci F^+, F^- to any point P on the ellipse takes the constant value $2a$, the length of the major axis.

This example is the basis of the *string construction* for the practical tracing of ellipses with given foci, a method long familiar to gardeners, groundsmen, graphic artists, and engineers. Mark two points F^+, F^- on a sheet of paper (for instance, by inserting drawing pins) and join them by a string of fixed length $2a$. Now take a pencil and place it against the taut string. The pencil tip P will move along a curve, which is the required ellipse. When F^+, F^- coincide, the curve will be the circle of radius a centred at that point.

Example 11.6 The focal properties of ellipses are of importance in astronomy. The orbit described by an object in space acting under an inverse square law

Table 11.1. *Eccentricities for the planets*

planet	e	planet	e	planet	e
Mercury	0.21	Saturn	0.06	Earth	0.02
Venus	0.01	Uranus	0.05	Neptune	0.01
Mars	0.09	Pluto	0.25	Jupiter	0.05

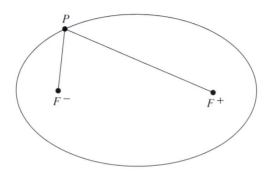

Fig. 11.3. The string construction

of attraction is known to be a conic, indeed a real ellipse, a hyperbola, or a parabola. For instance the planets describe elliptical orbits, with one focus at the sun. The eccentricities of the planets (correct to two decimal places) are listed in Table 11.1. It is noteworthy that they are small, so their orbits are very nearly circles: for instance, the foci of the earth's orbit are so close (in astronomical terms) that both lie inside the sun. Likewise, the orbits of communications satellites have extremely small eccentricities, so are virtually circular. By contrast, comets can have elliptical, hyperbolic, or parabolic orbits: for instance Halley's comet has an elliptical orbit of eccentricity ~ 0.97.

The reflective property of a parabola has an analogue for real ellipses, providing a good application of the natural trigonometric parametrization (4.5) of the ellipse.

Example 11.7 Consider the standard ellipse with moduli a, b and eccentricity e defined by $a^2 e^2 = a^2 - b^2$. Recall that the foci are the points $F^+ = (ae, 0)$, $F^- = (-ae, 0)$. We claim that for any point P on the ellipse the perpendicular bisectors of the lines L^+, L^- through F^+, P and F^-, P are the tangent and normal lines at P. It suffices to show that the tangent is parallel to a bisector. We can parametrize the ellipse as in (4.5) so $P = (ac, bs)$ with $c = \cos t$, $s = \sin t$. Then L^+, L^- are defined by

$$L^+ = bsx - a(c + e)y + abes, \qquad L^- = bsx - a(c - e)y - abes.$$

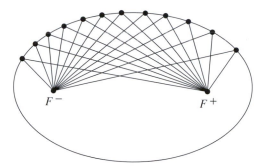

Fig. 11.4. Reflective property for an ellipse

Deleting the constant terms and multiplying out, we see that the parallel lines through the origin have joint equation

$$b^2 s^2 x^2 - 2abscxy + a^2(c^2 - e^2)y^2.$$

Direct computation verifies that their perpendicular bisectors have joint equation $(bcx + asy)(asx - bcy)$ and it remains to observe that by Example 9.7 the tangent line at P is the line $bcx + asy = ab$ parallel to the first factor.

Thus for any incident ray through one focus of an ellipse, the reflected ray passes through the other focus. Put another way, the pencil of rays emanating from one focus gives rise to a pencil of reflected rays through the other focus. The parabola can be thought of as a special case of the ellipse by thinking of its 'point at infinity' as a second 'focus': thus for any incident ray through the focus the reflected ray passes through the other 'focus'. Compared with the parabola, the reflective properties of the ellipse have received relatively little attention. In optics the property is used in designing car headlamps, to focus a light beam on the road ahead. And in acoustics it is used in the construction of 'whispering galleries', whereby one person standing at a focus can be heard whispering by a second person at the other focus, but by no-one else.

Exercise

11.3.1   A string of length 2 is attached to the points $(1, 0)$, $(0, 1)$. Show that the resulting ellipse is $3x^2 + 2xy + 3y^2 - 4x - 4y = 0$.

12

The Hyperbola

The geometry of the hyperbola has features in common with that of the real ellipse. Both types have a unique centre, two axes of symmetry, and two focal constructions. However, they differ fundamentally in one respect, namely that the hyperbola has two asymptotes. These represent a major feature of its geometry, providing the material for Section 12.1. The axes of a hyperbola are intimately related to the asymptotes, indeed they are their perpendicular bisectors. So far as parametrization is concerned, hyperbolas are analogous to ellipses. They cannot be parametrized by quadratic functions of a single variable, but do admit interesting rational parametrizations. For instance, the rectangular hyperbola has a rational parametrization very reminiscent of that for the circle. We use this to show how the geometry of the hyperbola underlies a standard technique of integration from foundational calculus.

12.1 Asymptotes

As we saw in Section 7.4 hyperbolas have two asymptotic directions, distinguishing them from ellipses and parabolas. Moreover, each asymptotic direction gives rise to a unique asymptote.

Lemma 12.1 *Any hyperbola H has exactly two asymptotes, namely the lines through the centre in the asymptotic directions.*

Proof Since $\delta < 0$ the quadratic terms in H factorize as UV, where U, V are lines through the origin in the asymptotic directions. Write $H = UV + W + c$, where W is linear and c is the constant term. By definition, asymptotes are parallel lines $L = U + p$, $M = V + q$ not intersecting H. We claim there are unique constants p, q, r with

$$H = LM + r = UV + (qU + pV) + (pq + r).$$

Since U, V are in different directions, the constants p, q are uniquely determined by $W = qU + pV$, and r by $c = pq + r$. Note that $r \neq 0$, else H is a line-pair LM: it follows that L, M cannot intersect H, so are asymptotes. Finally, observe that $H = LM + r$ has the same centre as the line-pair LM, namely the intersection of L, M. ☐

In an example one first determines the factors U, V of the quadratic terms in H: then p, q, r are found by equating coefficients of x, y and the constant term in $H = (U + p)(V + q) + r$.

Example 12.1 The hyperbola $H = p^2x^2 - q^2y^2 - 1$ has centre the origin. The quadratic terms factorize as the lines $px \pm qy$, which pass through the centre so must be the asymptotes.

Example 12.2 The reader will readily check that the conic H defined below is a hyperbola

$$H(x, y) = 77x^2 + 78xy - 27y^2 + 70x - 30y + 29.$$

The quadratic terms factorize as $(11x - 3y)(7x + 9y)$, so to determine the asymptotes we seek constants p, q, r for which

$$H(x, y) = (11x - 3y + p)(7x + 9y + q) + r.$$

Equating coefficients of x, y yields the following equations in p, q, r

$$7p + 11q = 70, \qquad 9p - 3q = -30, \qquad pq + r = 29.$$

The first two equations give $p = -1$, $q = 7$, and then the third yields $r = 36$. Thus the asymptotes are the lines

$$11x - 3y - 1 = 0, \qquad 7x + 9y + 7 = 0.$$

The asymptotes of a hyperbola H form a line-pair with vertex the centre, splitting the plane into two *asymptotic* cones. Consider now the way in which H intersects the pencil of all lines through its centre.

Lemma 12.2 *In one asymptotic cone every line through the centre meets H twice, and in the other no line meets H.*

Proof For a line L through the centre (u, v) in the direction (X, Y) parametrized as $x(t) = u + tX$, $y(t) = v + tY$ the intersection quadratic is as follows, where the coefficients are given by (4.3)

$$\phi(t) = pt^2 + qt + r = 0.$$

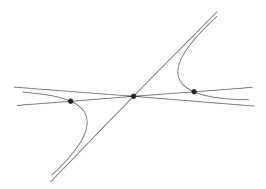

Fig. 12.1. A line in one asymptotic cone

As in the proof of Lemma 11.1 the coefficient $q = 0$, so intersections are given by the roots of $pt^2 + r = 0$. Moreover, r is non-zero and independent of the direction. Thus the quadratic has two distinct solutions when p, r have opposite signs, and no solutions otherwise. Further, p is given by the quadratic terms in H, so can be written $p = ST$, with S, T lines through the origin parallel to the asymptotes. But p has opposite signs on the two cones determined by S, T. Thus in one cone L does not intersect H, and in the other it intersects H twice. □

Thus the zero set of a hyperbola H splits into two *branches*, in opposite halves of one asymptotic cone, and symmetric under central reflexion in the centre. There are two lines through the centre of particular interest, namely the axes. The next result spells out their connexion with the asymptotes. As usual we assume H is given by a formula

$$H(x, y) = ax^2 + 2hxy + by^2 + 2gx + 2fy + c. \qquad (\star)$$

Lemma 12.3 *The axes of a hyperbola H are the perpendicular bisectors of its asymptotes. One axis meets H twice: the other does not meet H.*

Proof Lemmas 12.1 and 7.3 show that the asymptotes and axes all pass through the centre. Their directions are roots of the respective binary quadratics

$$aX^2 + 2hXY + bY^2 = 0, \qquad hX^2 + (b - a)XY - hY^2 = 0.$$

It follows that the axes are the perpendicular bisectors of the asymptotes. As we pointed out in Section 6.4, the bisectors lie in different cones: so one axis meets H twice, whilst the other does not meet H. □

The axis of a hyperbola H intersecting it twice is the *transverse axis*; and the axis not intersecting H is the *conjugate* axis. These definitions are consistent with those given for the standard conics in Section 4.1. Thus a hyperbola has two distinct vertices, whose distance apart is the *length* of the transverse axis.

Example 12.3 The hyperbola $H = p^2x^2 - q^2y^2 - 1$ has axes $x = 0$, $y = 0$: the former does not meet H (so is the conjugate axis), whilst the latter meets H at the points where $x = \pm 1/p$ (so is the transverse axis). The standard hyperbola with moduli a, b is the case when $p = 1/a$, $q = 1/b$, with vertices the points $(\pm a, 0)$.

The point of the next example is that it illustrates how elementary geometry can throw light on a topic in foundational calculus, namely that of sketching graphs of rational functions $y = g(x)/h(x)$. Any value $x = s$ with $h(s) = 0$ gives rise to a line $x = s$ which the graph cannot intersect. When g is quadratic and h is linear, the graph is a hyperbola, and the geometric significance of the line $x = s$ is that it is an asymptote. Calculus methods do not produce the second asymptote: its relevance to the graph is only clear from the geometry.

Example 12.4 Consider the graph of a rational function $y = g(x)/h(x)$, where g, h are given by formulas of the form

$$g(x) = px^2 + 2qx + r, \qquad h(x) = x - s.$$

We assume $h(x)$ is not a factor of $g(x)$, so $g(s) \neq 0$. A moment's thought will show that the graph is the zero set of the conic H defined by

$$H(x, y) = yh(x) - g(x) = -px^2 + xy - 2qx - sy - pr.$$

The conic H is a hyperbola, with unique centre $x = s$, $y = 2(ps + q)$. Its quadratic terms are $x(y - px)$, so the asymptotes are the lines through the centre parallel to $x = 0$, $y = px$, namely $x = s$, $y = p(x + s) + 2q$. Figure 12.2 illustrates the case $p = 1$, $q = -1$, $r = 2$, $s = 1$ with asymptotes the lines $x = 1$, $y = x - 1$.

Example 12.5 The tangent to the hyperbola $xy = 1$ at the point (u, v) is the line $vx + uy = 2uv$. Parametrize the branch of the hyperbola in the first quadrant as $x(t) = t$, $y(t) = 1/t$ with $t > 0$. Then the tangent at t is $x + t^2y - 2t$, with direction $-t^2 : 1$. The limiting tangent direction as $t \to \infty$ is $1 : 0$, with limiting tangent $y = 0$: likewise, the limiting tangent direction as

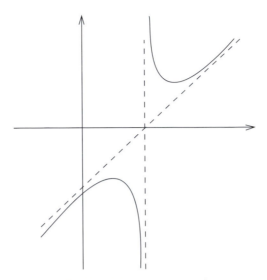

Fig. 12.2. A hyperbola as a graph

$t \rightarrow 0$ is $0 : 1$, with limiting tangent $x = 0$. This example suggests that the asymptotes $x = 0$ and $y = 0$ can be regarded as 'tangents' to the hyperbola at the points 'at infinity' in the directions $0 : 1$ and $1 : 0$.

In fact this idea can be formalized. In *EGAC* it is shown that any conic can be considered in the projective plane, obtained from the ordinary plane by adding 'points at infinity'. Tangents can then be introduced analogously, and asymptotes appear naturally as tangents to the conic at 'points at infinity'.

Exercises

12.1.1   In each of the following cases verify that the given conic is a hyperbola, and find its centre, axes and asymptotes:

(i) $6xy + 9x + 4y = 0$,
(ii) $2x^2 - xy - 3y^2 + 4x - 1 = 0$,
(iii) $3x^2 + 6xy + 2y^2 + 6x + 10y + 1 = 0$,
(iv) $x^2 + 2xy - 3y^2 + 8y + 1 = 0$.

12.1.2   In each of the following cases verify that the given conic is a hyperbola, and find its centre, axes, and asymptotes:

(i) $4x^2 - 9y^2 - 24x - 36y - 36 = 0$,
(ii) $6x^2 + 11xy - 10y^2 - 4x + 9y = 0$,
(iii) $y^2 - 4xy - 5x^2 + 42x + 6y - 63 = 0$.

12.1.3 Find the hyperbola which touches the y-axis at the origin, touches the line $7x - y - 5 = 0$ at the point $(1, 2)$, and has an asymptote parallel to the x-axis.

12.2 Parametrizing Hyperbolas

Recall again the natural parametrization of parabolas by quadratic functions of the form

$$x(t) = x_0 + x_1 t + x_2 t^2, \qquad y(t) = y_0 + y_1 t + y_2 t^2. \qquad (12.1)$$

In the previous chapter we noted that such parametrizations cannot be extended to ellipses. Likewise, it turns out that hyperbolas do not admit parametrizations of this form. There is however a halfway house. The derivation of (12.1) used the fact that the axis direction for a parabola is asymptotic, so lines parallel to the axis meet the parabola at a unique point. Since a hyperbola H has two asymptotic directions it is natural to consider its intersection with the parallel pencils of lines in those directions. To implement this idea, write $H = LM + r$ where L, M are the asymptotes, and r is a non-zero constant. Now consider the intersection of H with the pencil of parallel lines $L = t$. Since L itself does not intersect H, every point on H lies on just one line $L = t$, with $t \neq 0$. The point of intersection is determined by

$$L(x, y) = t, \qquad M(x, y) = -\frac{r}{t}. \qquad (12.2)$$

Each non-zero value of t corresponds to a unique point on H. For positive t the corresponding points lie on one branch, and for negative t on the other branch. Solving the two linear equations (12.2) for x, y explicitly in terms of t we obtain solutions of the following form, parametrizing one branch when $t > 0$, and the other when $t < 0$

$$x(t) = \frac{x_0 + x_1 t + x_2 t^2}{t}, \qquad y(t) = \frac{y_0 + y_1 t + y_2 t^2}{t}. \qquad (12.3)$$

Example 12.6 A simple illustration is provided by the hyperbola $H = xy - 1$. In this case $L(x, y) = x$, $M(x, y) = y$, $r = -1$ and (12.2) have the solutions $x(t) = t$, $y(t) = 1/t$.

Example 12.7 The hyperbola $H = 6x^2 + 11xy - 10y^2 - 4x + 9y = 0$ can be written $H = LM + r$, where $L = 3x - 2y + 1$, $M = 2x + 5y - 2$ and $r = 2$. Solving the equations (12.2) we obtain the parametrization

$$x(t) = \frac{(t - 1)(5t + 4)}{19t}, \qquad y(t) = \frac{-2(t - 1)(t - 3)}{19t}.$$

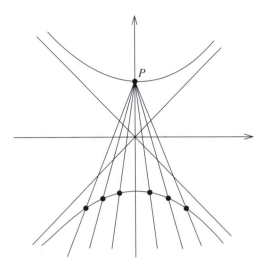

Fig. 12.3. Parametrizing a rectangular hyperbola

Observe that (12.3) are *rational* parametrizations of the hyperbola H. By considering the pencil of lines through a fixed point P on H one obtains rational parametrizations akin to that obtained for the circle.

Example 12.8 Consider the point $P = (0, a)$ on the 'upper' branch of the rectangular hyperbola $H = y^2 - x^2 - a^2$. Any 'non-vertical' line through P has the form $y = tx + a$ for some t and meets H at P, and some other point whose coordinates depend on t. To find the second point, substitute $y = tx + a$ in $H(x, y) = 0$ to obtain the quadratic

$$x\{(t^2 - 1)x + 2at\} = 0.$$

The root $x = 0$ corresponds to the intersection with H at P itself. The other root gives the required rational parametrization

$$x(t) = \frac{2at}{1 - t^2}, \qquad y(t) = tx(t) + a = a\left(\frac{1 + t^2}{1 - t^2}\right).$$

For $-1 < t < 1$ these formulas parametrize the 'upper' branch, whilst for $t < -1$ and $t > 1$ they parametrize the two halves of the 'lower' branch, missing the single point $(0, -a)$. Note that the formulas are undefined at $t = \pm 1$, corresponding to the lines through P in the asymptotic directions.

Here is an application. It makes sense of a calculus technique for integrating certain functions arising in the physical sciences, where it is traditional to suppress the underlying geometric idea.

Example 12.9 The usual technique for evaluating the indefinite integral below is on the level of a trick

$$\int \frac{dx}{\sqrt{1+x^2}}.$$

The technique is to make the substitution $x = 2t/(1-t^2)$, reducing the integral to one of the following form

$$\int \frac{2\,dt}{1-t^2} = \int \frac{dt}{1+t} + \int \frac{dt}{1-t} = \log\left(\frac{1+t}{1-t}\right).$$

Clever though this may be, it begs a question. Is there a way of viewing the problem in which the substitution is a natural step? The key is to write $y = \sqrt{1+x^2}$, and observe that then (x, y) is a point on the 'upper' branch of the hyperbola $y^2 - x^2 = 1$. As in Example 12.8 that branch is parametrized by the formulas below, with $-1 < t < 1$, and we see that the substitution is a perfectly natural one

$$x(t) = \frac{2t}{1-t^2}, \qquad y(t) = \frac{1+t^2}{1-t^2}.$$

Exercises

12.2.1   Show that the hyperbola $x^2 + 2xy - 3y^2 + 8y + 1 = 0$ can be parametrized by the formulas

$$x(t) = \frac{-3t^2 - 4t + 5}{t}, \qquad y(t) = \frac{t^2 + 4t + 5}{t}.$$

12.2.2   By considering the pencil of lines through the point $P = (-a, 0)$ show that the standard rectangular hyperbola $x^2 - y^2 = a^2$ has the rational parametrization

$$x(t) = a\left(\frac{1+t^2}{1-t^2}\right), \qquad y(t) = \frac{2at}{1-t^2}.$$

12.3 Focal Properties of Hyperbolas

Virtually no mention has been made of the trace invariant τ since its introduction in Chapter 4. It is however of particular relevance to hyperbolas, for the following reason. Consider the general conic

$$H(x, y) = ax^2 + 2hxy + by^2 + 2gx + 2fy + c. \qquad (\star)$$

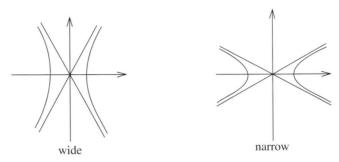

wide   narrow

Fig. 12.4. Wide and narrow hyperbolas

We ask when the trace invariant can vanish. Recall that τ is defined by the formula $\tau = a + b$. Thus $\tau = 0$ if and only if $b = -a$. In that case the delta invariant is $\delta = -a^2 - h^2 < 0$. Thus for a non-degenerate conic H the trace invariant can only vanish when H is a hyperbola.

Example 12.10 For the standard hyperbola with moduli a, b we can interpret the vanishing of the trace invariant in terms of the eccentricity e. For the standard hyperbola $\tau = 0$ if and only if $a = b$. By Lemma 8.5 the eccentricity is defined by the relation $a^2 e^2 = a^2 + b^2$: thus $a = b$ if and only if $e^2 = 2$. On this basis a general hyperbola is said to be *rectangular* when $e = \sqrt{2}$. For standard hyperbolas the rectangular type can be regarded as a transitional type between standard hyperbolas with $a > b$ (having 'wide' branches) and those with $b > a$ (having 'narrow' branches). As we will see in Section 13.3, these subclasses of hyperbolas exhibit a surprising geometric difference.

In Example 11.5 we showed that the *sum* of the distances from a variable point P on an ellipse to its foci is constant. There is an analogous result for hyperbolas, namely that on each branch the *difference* of the distances from a variable point P to the foci is constant.

Example 12.11 Let H be a standard hyperbola with moduli a, b and eccentricity e. The foci are denoted F^+, F^- and the directrices D^+, D^- so for any point P on the hyperbola we have $PF^+ = ePD^+$, $PF^- = ePD^-$. We then have

$$PF^+ - PF^- = e(PD^+ - PD^-) = e\left(\frac{\pm 2a}{e}\right) = \pm 2a.$$

The '$+$' case arises when P is on the negative branch of H, and the '$-$' case when P is on the positive branch. (Figure 12.5.) Thus in absolute value, the

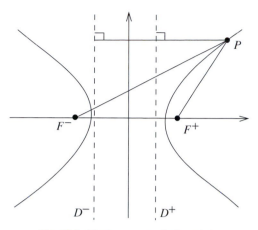

Fig. 12.5. Metric property of a hyperbola

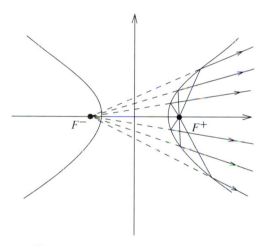

Fig. 12.6. Reflective property for a hyperbola

difference between the distances from P to the foci F^+, F^- has the constant value $2a$, the length of the transverse axis.

Example 12.12 The hyperbola shares a reflective property of the ellipse, described in Example 11.7, namely that for any incident ray emanating from one focus, the reflected ray passes through the other focus. Consider the standard hyperbola H with moduli a, b, eccentricity e, and foci the points $F^+ = (ae, 0)$, $F^- = (-ae, 0)$. For a general point P on H we write L^+, L^- for the lines through F^+, P and F^-, P. Then the claim is that the perpendicular bisectors of L^+, L^- are the tangent and normal lines to H at P. The proof is virtually identical to that given for the ellipse. We parametrize H as in (4.6) so $P = (\pm a \cosh t, b \sinh t)$, the '+' case for the positive branch, and the '−' case for the negative branch. We leave the reader to repeat the details of Example 11.7, replacing the trigonometric functions by the hyperbolic functions.

The reflective property of the hyperbola (like that of the ellipse) has received little attention compared with that given to the parabola. In optics it has been used to construct compact reflecting telescopes, using a hyperbolic secondary mirror.

13

Pole and Polar

Understanding the way in which a conic Q intersects lines is a recurring theme in this text. The foundations were laid in Chapter 4, where we studied intersections of Q with a single line. The idea was developed in Chapter 7 by studying intersections with parallel pencils. And Chapter 9 continued the theme, by studying the intersections of Q with the pencil of lines through one of its points. In this chapter we remove all the restrictions, by considering the intersections of Q with the pencil of all lines through an *arbitrary* point W in the plane.

The case when Q is a circle suggests that the qualitative picture depends on how W lies relative to Q. (Figure 13.1.) The transitional case, when W lies on Q, was elucidated in the previous chapter: every line through W meets Q once again, with the sole exception of the tangent line for which the intersections coalesce. When W is *inside* Q we expect every line through W to intersect Q twice. But when W is *outside* Q we expect some lines to intersect Q twice, some lines not to intersect Q at all, and just two exceptional lines to be tangent.

13.1 The Polars of a Conic

We could verify our expectations for a circle by direct calculation. It is however more productive to consider the question for a general conic

$$Q(x, y) = ax^2 + 2hxy + by^2 + 2gx + 2fy + c. \qquad (\star)$$

We need an efficient technique to determine the tangents to Q passing through a given point. To motivate the definition (in some measure) recall that (17.2) expresses Q in terms of its matrix A by the following formula, where $z = (x, y, 1)$

$$Q(x, y) = zAz^T.$$

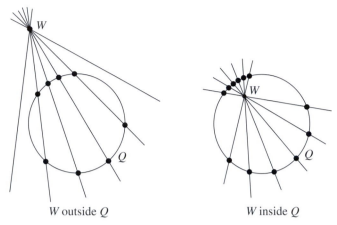

W outside Q   W inside Q

Fig. 13.1. Circle intersecting pencils of lines

Now let $P = (u, v)$ be any point. We define the *polar* of Q with respect to P to be the following function, where $w = (u, v, 1)$

$$L(x, y) = wAz^T. \tag{13.1}$$

It will be useful to express this in more concrete terms. Carrying out the matrix multiplications we find that the polar is given explicitly by the formula

$$L(x, y) = x(au + hv + g) + y(hu + bv + f) + (gu + fv + c).$$

Example 13.1 The formula shows that the polar of the unit circle $Q = x^2 + y^2 - 1$ with respect to $P = (u, v)$ is $L = ux + vy - 1$. More generally, for a conic $Q = \alpha x^2 + \beta y^2 + \gamma$ we have $L = \alpha ux + \beta vy + \gamma$. (That covers the standard circles, ellipses, and hyperbolas.) Observe that L is a line, save when P is the centre, and $u = 0$, $v = 0$.

Some of the more basic properties of the polar can be deduced by writing it in the the form

$$2L(x, y) = 2Q(u, v) + (x - u)Q_x(u, v) + (y - v)Q_y(u, v). \tag{13.2}$$

The coefficients of x, y are $Q_x(u, v)$, $Q_y(u, v)$. Provided these do not both vanish, L defines a line, the *polar line* of Q with respect to the *pole P*. The function L only fails to be a line when both partials vanish, which according to Lemma 5.3 is the condition for P to be a centre for Q. A special case arises when the pole lies on Q: in that case $Q(u, v) = 0$ so by (9.1) the polar line is the tangent line to Q at P.

Exercises

13.1.1 Show that the polar lines of the points $(1, 3)$, $(2, 1)$, $(3, -1)$ with respect to the circle $x^2 + y^2 = 4$ are concurrent.

13.1.2 Let Q be a conic, and let $W = (u, v)$, $Z = (x, y)$ be points. Show that Z lies on the polar line with respect to the pole W if and only if W lies on the polar line with respect to the pole Z.

13.1.3 Let M be a line. Show that the polar line of the repeated line $Q = M^2$ with respect to a pole (u, v) not on M is $M(u, v)M$, and deduce that J is identically zero.

13.1.4 Use the standard parametrization of the circle $(x - a)^2 + y^2 = 4a^2$ to show that the polar of any point on that circle with respect to the circle $(x + a)^2 + y^2 = 4a^2$ touches the parabola $y^2 = -4ax$.

13.1.5 Let C, C' be circles having distinct centres. Show that if the polar lines of a point A with respect to C, C' intersect at B then the perpendicular bisector of the line joining A, B is the radical axis of the circles.

13.1.6 Let P be a point on the hyperbola $2y^2 - x^2 = 1$. Show that the polar line of P with respect to the parabola $y^2 = x$ touches the hyperbola.

13.2 The Joint Tangent Equation

The geometric significance of the polar line is explained by the following result.

Lemma 13.1 *Let $P = (u, v)$ be a point which is not a centre of the conic Q, and let L be the polar line with respect to P. Then a point (x, y) lies on a line through P tangent to Q if and only if $J(x, y) = 0$, where*

$$J(x, y) = L(x, y)^2 - Q(u, v)Q(x, y). \tag{13.3}$$

Proof The line through (u, v), (x, y) is parametrized as $x(t) = u + tX$, $y(t) = v + tY$ where $X = x - u$, $Y = y - v$ and meets Q when the intersection quadratic

$$\phi(t) = Q(u + tX, v + tY) = pt^2 + qt + r = 0.$$

The line is tangent to Q if and only if this quadratic has a repeated root, i.e. if and only if its discriminant $q^2 - 4pr = 0$. It is convenient to write this in the equivalent form

$$(q + 2r)^2 = 4r(p + q + r).$$

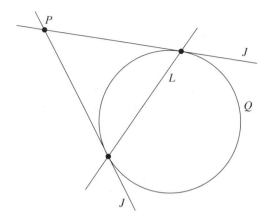

Fig. 13.2. Pole and polar

It remains to identity this relation with (13.3). First, observe that

$$r = \phi(0) = Q(u, v), \qquad p + q + r = \phi(1) = Q(x, y).$$

It remains to note that the formulas (4.3) for the coefficients in ϕ give the following expression for q: combining this with (13.2) gives $q + 2r = 2L$ as required

$$q = (x - u)Q_x(u, v) + (y - v)Q_y(u, v).$$

$\square$

Here is the mental picture. In principle J is a conic. Suppose P does not lie on Q, and the polar L is a line, intersecting Q in two points. The formula (13.3) shows that the intersections of L, Q coincide with those of J, Q. Thus there are two points on Q which lie on tangents to Q through P. Indeed, those points must be the points of tangency, and L is the line joining them. Moreover, every point on each tangent through P lies on Q, so the tangents are the components of J. For that reason J is called the *joint tangent equation* of Q with respect to P. In the special case when P lies on Q we have $Q(u, v) = 0$ so L is the tangent at P, and $J = L^2$ is the repeated tangent line.

Example 13.2 For the unit circle $C = x^2 + y^2 - 1$ the polar line of C with respect to a point $P = (u, v)$ distinct from the centre is the line $L = 2(ux + vy - 1)$. And the joint equation of the tangents is

$$(ux + vy - 1)^2 = (u^2 + v^2 - 1)(x^2 + y^2 - 1).$$

For instance, when $P = (0, \sqrt{2})$ the joint equation of the tangents is the real line-pair displayed below

$$x^2 - y^2 + 2\sqrt{2}y - 2 = (x - y + \sqrt{2})(x + y - \sqrt{2}).$$

The tangents through P are then the components of the joint equation, i.e. the lines $x - y + \sqrt{2} = 0$, $x + y - \sqrt{2} = 0$. Note that the tangents in this example are perpendicular, since the coefficients of x^2, y^2 in the joint equation are respectively 1, -1, with zero sum.

Example 13.3 The tangent line at a point (a, b) on the rectangular hyperbola $Q(x, y) = x^2 - y^2 - 1$ has equation $ax - by = 1$. It follows that no tangents to Q pass through the origin. That appears to be at odds with the fact that when the pole is the centre $u = 0$, $v = 0$ the joint equation $J(x, y) = x^2 - y^2$ *does* represent a real line-pair, in fact the asymptotes $y = \pm x$. What is happening here is that there are points (u, v) arbitrarily close to the origin where tangents to Q intersect, and have the asymptotes as limiting positions. Of course Lemma 13.1 does not apply, since when $u = 0$, $v = 0$ the pole is a centre.

The fact that no tangents pass through the centre in this example is a special case of a visually compelling general fact.

Example 13.4 Generally speaking, if the centre P of a conic Q does not lie on Q then no tangent to Q can pass through it. Indeed for any point Z on Q there is a corresponding point Z' on Q for which Z, P, Z' are collinear. Were the tangent at Z to pass through P it would also pass through Z'. In that case the tangent would meet Q in at least three distinct points, so would be a component of Q by the Component Lemma: that contradicts the assumption that P does not lie on Q.

It is natural to ask for the significance of the polar line when it fails to intersect Q. The most satisfying answer (mathematically) is reserved for readers who pursue the geometry of *complex* conics, for instance in *EGAC*. However, even within the confines of this text, such a situation is not without interest.

Example 13.5 The polar line with respect to the pole (u, v) for the standard parabolas $Q = y^2 - 4ax$ with $a > 0$ is

$$L(x, y) = vy - 2a(x + u).$$

In particular, the polar line with respect to a pole $(u, 0)$ on the x-axis is the line $x = -u$. For $u > 0$ the polar line fails to intersect Q; for $u = 0$ it is the tangent line to Q at its vertex; and for $u < 0$ it intersects Q in two distinct points, the tangents at those points intersecting at the pole. The interest in this example is that in the case $u = a$, when the pole is the focus, the polar line is the directrix $x = -a$.

Naturally, we would like to know how the joint tangent equation J fits into our broad subdivisions of conics into types. The next result provides us with useful information. We assume henceforth that Q is not a repeated line, the only class of conic for which J is identically zero, for any choice of pole. (Exercise 13.1.3.)

Lemma 13.2 *Let $P = (u, v)$ be a point which is not a centre for Q, and let J be the joint tangent equation of Q with respect to P. Then P is singular on J.*

Proof It follows immediately from (13.2) that $L(u, v) = Q(u, v)$, where L is the polar of Q with respect to P. Then P lies on J since

$$J(u, v) = L^2(u, v) - Q^2(u, v) = 0.$$

It remains to show (u, v) is a centre of J. Differentiating J with respect to x, y, writing $x = u$, $y = v$ in the result, and using the fact that $L(u, v) = Q(u, v)$, we obtain

$$\begin{cases} J_x(u, v) = L(u, v)\{2L_x(u, v) - Q_x(u, v)\} \\ J_y(u, v) = L(u, v)\{2L_y(u, v) - Q_y(u, v)\}. \end{cases}$$

However, differentiating (13.2) with respect to x, y, and setting $x = u$, $y = v$ in the result, we obtain the relations

$$2L_x(u, v) = Q_x(u, v), \qquad 2L_y(u, v) = Q_y(u, v).$$

It follows immediately that J_x, J_y both vanish at P, and hence that P is a centre for J. $\qquad\square$

In Example 6.1 we pointed out that singular conics are automatically degenerate. In particular, any joint tangent equation J of a conic Q is a degenerate conic, and is therefore a real, parallel, or virtual line-pair. (Table 6.1.)

Example 13.6 As we saw above, the joint equation of the tangents to the unit circle $C = x^2 + y^2 - 1$ with respect to the pole $P = (u, v)$ is given by

$$(ux + vy - 1)^2 = (u^2 + v^2 - 1)(x^2 + y^2 - 1).$$

To determine the type of this degenerate conic, observe that the quadratic terms are

$$(1 - v^2)x^2 + 2uvxy + (1 - u^2)y^2.$$

It follows that the delta invariant is $\delta = 1 - u^2 - v^2$. When P is inside C we have $\delta > 0$, and the joint equation is a virtual line-pair. When P is on C we have $\delta = 0$ and the joint equation is a parallel line-pair, indeed the repeated tangent line $(ux + vy - 1)^2 = 0$. And when P is outside C we have $\delta < 0$, and the joint equation is a real line-pair.

That confirms our expectations for the circle set out in the introduction to this chapter. We can pursue such examples further, to determine the tangents through a given point, using the method of Section 6.2 to factorize their joint equation.

Exercises

13.2.1 In each of the following cases find the tangents to Q through P and decide whether or not they are perpendicular:

(i) $Q = x^2 + y^2 - 25$, $P = (-1, 7)$,
(ii) $Q = x^2 + y^2 - 6x - 10y + 25$, $P = (0, 0)$,
(iii) $Q = x^2 + y^2 + 2x - 14y + 18$, $P = (0, 0)$,
(iv) $Q = x^2 + y^2 - 14x + 2y + 25$, $P = (0, 0)$.

13.2.2 Find the two points on the x-axis from which the tangents to the circle C below are perpendicular, and the tangents in each case

$$C(x, y) = x^2 + y^2 - 10x - 8y + 31.$$

13.2.3 In each of the following cases show that the joint tangent equation with respect to a non-central pole (u, v) is a repeated line:

(i) $Q(x, y) = x^2 - y^2$,   (iii) $Q(x, y) = x^2 - 1$,
(ii) $Q(x, y) = x^2 + y^2$,   (iv) $Q(x, y) = x^2 + 1$.

13.2.4 Let Q be the standard ellipse with moduli a, b. Show that the points P for which the tangents to Q through P have a fixed line $y = x \tan \theta$ as a bisector lie on the rectangular hyperbola

$$x^2 - 2xy \cot \theta - y^2 = a^2 - b^2.$$

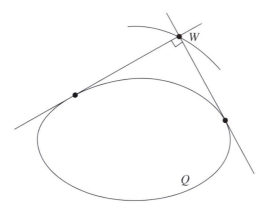

Fig. 13.3. The idea of the orthoptic locus

13.3 Orthoptic Loci

Here is an interesting geometric idea, which arises in studying the irregularities of the human eye. The *orthoptic locus* of a non-degenerate conic Q is the set of points W with the property that there are two perpendicular tangents to Q through W. For a good mental picture, think of Q as a shape cut out of thin rigid material, and fixed on a plane surface. (Figure 13.3.) Now take a carpenter's set square, representing a pair of perpendicular lines, and place it so that it touches Q at two points. Then the orthoptic locus of Q is the locus of the apex as the set square moves around Q.

Example 13.7 The simplest possible instance of an orthoptic locus is the unit circle $x^2 + y^2 = 1$. It is easy enough to spot a point on the locus. For instance, the lines $x = 1$, $y = 1$ are perpendicular tangents, intersecting at a point distant $\sqrt{2}$ from the centre. The rotational symmetry of the circle then suggests that the orthoptic locus will comprise all points at distance $\sqrt{2}$ from the centre, so should be the concentric circle $x^2 + y^2 = 2$ of radius $\sqrt{2}$.

By contrast, it is not so obvious *a priori* what the orthoptic locus of the standard conics will be. Lemma 13.1 provides the basis for an efficient technique. Recall that by Example 6.5 two tangents to Q through a point (u, v) are perpendicular if and only if the sum of the coefficients of x^2, y^2 in their joint equation is zero. That gives an equation in u, v which must be satisfied by any point in the orthoptic locus. However, the reader should beware that a point (u, v) satisfying this equation is only in the orthoptic locus if there are indeed two tangents to Q through it.

Example 13.8 We will illustrate the technique described above for the standard ellipses and circles

$$Q(x, y) = \frac{x^2}{a^2} + \frac{y^2}{b^2} - 1, \qquad (0 < b \le a).$$

By Lemma 13.1, the joint equation of the tangents through an arbitrary point (u, v) is

$$J(x, y) = \left(\frac{ux}{a^2} + \frac{vy}{b^2} - 1\right)^2 - \left(\frac{u^2}{a^2} + \frac{v^2}{b^2} - 1\right)\left(\frac{x^2}{a^2} + \frac{y^2}{b^2} - 1\right).$$

By inspection, we see that the coefficients A, B of x^2, y^2 in this expression are given by

$$A = \frac{b^2 - v^2}{a^2 b^2}, \qquad B = \frac{a^2 - u^2}{a^2 b^2}.$$

The condition for the tangents through (u, v) to be perpendicular is that the sum of these two expressions should vanish. Thus the orthoptic locus of a standard circle or ellipse Q lies on a concentric circle

$$u^2 + v^2 = a^2 + b^2.$$

Conversely, every point (u, v) on this circle is in the orthoptic locus. We need only check that through such points there are two tangents to Q. It suffices to show that for such points the delta invariant of J is negative, so J is a real line-pair. The calculation is left to Exercise 13.3.1.

The simplest possible instance of this example is the case $a = b = 1$ of the unit circle C, when we confirm the conclusion of the previous example, namely that the orthoptic locus is the concentric circle of radius $\sqrt{2}$.

Example 13.9 The orthoptic locus of the hyperbola is more puzzling than the ellipse. The reader is invited to repeat the computations of the previous example, replacing the standard ellipses of that example by standard hyperbolas

$$Q(x, y) = \frac{x^2}{a^2} - \frac{y^2}{b^2} - 1, \qquad (a, b > 0).$$

The net result of the computation is that any point (u, v) in the orthoptic locus lies on the following conic, defining a virtual circle when $a < b$, a point circle when $a = b$, and a real circle when $a > b$

$$u^2 + v^2 = a^2 - b^2. \qquad (13.4)$$

In particular, in the case $a < b$ when the hyperbola has wide branches the zero

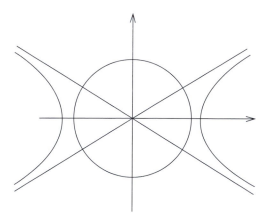

Fig. 13.4. Orthoptic locus of a narrow hyperbola

set of the conic (13.4) is empty, and hence the orthoptic locus is empty: the mental picture is that any two tangents intersect at an obtuse angle. The case $a = b$ of the rectangular hyperbola is misleading: the zero set of the point circle is the centre of the hyperbola, and is *not* in the orthoptic locus, for the reason given in Example 13.3: thus the locus is again empty. Even in the case $a > b$ when the hyperbola has narrow branches one has to be careful. In that case (13.4) is a real circle intersecting the transverse axis between the vertices. However, there are four points on the circle, namely its intersections with the asymptotes $bx \pm ay = 0$, which fail to be in the orthoptic locus.

For the sake of completeness we will also determine the orthoptic locus of a parabola, which is of relevance to the material in Chapter 8.

Example 13.10 For a standard parabola $C(x, y) = y^2 - 4ax$ with $a > 0$ the polar line with respect to the pole (u, v) is $vy - 2ax - 2au$. Thus the joint equation of the tangents is

$$J(x, y) = (vy - 2ax - 2au)^2 - (v^2 - 4au)(y^2 - 4ax).$$

The coefficients of x^2, y^2 in J are $4a^2$, $4au$, whose sum is zero if and only if $u = -a$. It follows that the orthoptic locus of a standard parabola is the directrix $u = -a$.

By combining the result of this example with the listing of conics in Chapter 15 it turns out that the orthoptic locus of *any* parabola is its directrix line D. That yields an alternative method for finding D, and hence the focus F. First find the axis M and directrix D; then find the vertex V and the intersection W

of M with D. Since V is the midpoint of the axis segment joining F, W, the focus is given by $F = 2V - W$.

Example 13.11 In Example 10.3 we analyzed the geometry of the parabola defined by

$$Q(x, y) = 4x^2 - 4xy + y^2 - 10y - 19.$$

The axis is the line $M = 2x - y + 1$, the perpendicular line through the vertex is $M' = x + 2y + 5$, and the vertex is

$$V = \left(-\frac{7}{5}, -\frac{9}{5}\right).$$

The orthoptic locus can be derived using the general method of Section 13.3. The coefficients of x^2, y^2 in the joint tangent equation $J(x, y)$ are readily checked to be $4(10v + 19)$, $4(5u + 11)$, and the locus is given by the vanishing of the sum, so is the line $u + 2v + 6 = 0$. That agrees with the result given by the formula (13.5). Moreover, it is consistent with the geometry, since the directrix should be parallel to M'. The intersection W of the directrix with the axis, and the focus F, are then given by

$$W = \left(-\frac{8}{5}, -\frac{11}{5}\right), \qquad F = 2V - W = \left(-\frac{6}{5}, -\frac{7}{5}\right).$$

Finally, it is worth mentioning that one can mechanize the derivation of the orthoptic locus, by applying the above method to a general conic $(\star)$. There is of course a computational advantage in having a simple formula applicable to any conic. However, it also helps to clarify the geometry. The enterprising reader (armed with paper, pen, and a degree of patience) might like to verify that, using the notation of (5.2), the equation of the orthoptic locus of $(\star)$ is

$$C(x^2 + y^2) - 2Gx - 2Fy + (A + B) = 0. \qquad (13.5)$$

That defines a *conic* if and only if $C = \delta \neq 0$. In that case $(\star)$ has a unique centre, and the locus is a circle. Moreover, the centre of the orthoptic locus has coordinates

$$u = \frac{G}{C}, \qquad v = \frac{F}{C}.$$

By (5.3) that coincides with the centre of $(\star)$. Thus the orthoptic locus is concentric with $(\star)$, as we discovered for the standard ellipses and hyperbolas. Finally, when $\delta = 0$ the relation (13.5) defines a line: in particular, the orthoptic locus of a parabola is always a line, as we claimed above.

Exercises

13.3.1   Example 13.8 studied the orthoptic locus arising from the standard ellipses and circles with moduli a, b satisfying $0 < b \le a$. Show that the delta invariant of the joint equation J of the tangents through (u, v) in that example is

$$\delta = \frac{-1}{a^2 b^2} \left\{ \frac{u^2}{a^2} + \frac{v^2}{b^2} - 1 \right\}.$$

Use this formula to show that $\delta < 0$ for points (u, v) on the concentric circle displayed below, and deduce that this circle is the orthoptic locus

$$u^2 + v^2 = a^2 + b^2.$$

13.3.2   Supply the missing detail in Example 13.9. Show that any point (u, v) in the orthoptic locus of a standard hyperbola with moduli a, b satisfies the equation $u^2 + v^2 = a^2 - b^2$. In the case $a > b$ verify that every point on this real circle lies in the orthoptic locus, with four exceptions.

13.3.3   Let a, b be non-zero constants. Show that the zero set of the conic Q_λ defined below is non-empty

$$Q_\lambda(x, y) = (ax + by - 1)^2 - 2\lambda xy.$$

Calculate the invariants of Q_λ, and hence determine its type in terms of the parameter λ. Find the orthoptic locus of Q_λ in the case when it is non-degenerate.

14

Congruences

An interesting facet of human psychology is that we perceive differences between objects more readily than we do similarities. For instance, looking at the illustrations of standard conics in Chapter 4 you probably feel there are clear differences between an ellipse and a hyperbola, but may be less sure about the similarities between two parabolas. In mathematics we can only assert that two objects are 'different' when we are perfectly clear about what we mean by them being 'the same'. Formalising this idea is the immediate problem facing us. One of the sublimal messages of geometry is that there is no one answer, it all depends on your objectives. People who take a 'black or white' view of the world may find this unsettling, whereas those who relish the gamut of intermediate greys will sense interesting possibilities.

We adopt the approach of greatest relevance in the physical sciences. In a nutshell, the idea is to think of two conics as being 'the same' when the one can be superimposed on the other. That is an eminently practical criterion. Suppose for instance that you have two ellipses drawn on a flat surface. Each ellipse could be traced on to a plastic transparency with a felt tip pen. Mark the centres of the ellipses. We could proceed in two steps. First we could slide one transparency across the surface till the centres of the two ellipses coincide: then we could rotate the transparency about the common centre to see if they superimpose.

Thus we expect 'translation' and 'rotation' of the plane to be sufficient for our purposes. Both concepts are special cases of the general 'congruences' introduced in Section 14.1. The next two sections systematically study the effect of congruences on lines and conics, with the broad objective of showing that congruences leave the geometry invariant. Section 14.4 goes one step further by stating the Invariance Theorem, that the basic invariants of conics are left invariant by congruences. (Incidentally, that is the justification for the

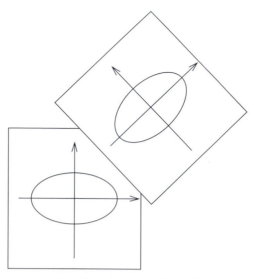

Fig. 14.1. Superimposition of two ellipses

nomenclature.) The proof is technical, and for that reason postponed to
Chapter 17 for the interested reader.

14.1 Congruences

We need some preliminary comments about planar maps ϕ. Think of the domain as the (X, Y)-plane, and the target as the (x, y)-plane, so ϕ is given by a
formula of the following form

$$\phi(X, Y) = (x, y). \tag{14.1}$$

In this formula x, y are abbreviations for two functions $x = x(X, Y)$, $y = y(X, Y)$, the *components* of ϕ. Recall that ϕ is *invertible* when for any point
(x, y) in the target there exists a unique point (X, Y) in the domain for which
(14.1) holds. Solving this equation for X, Y in terms of x, y we obtain functions $X(x, y)$, $Y(x, y)$ which are the components of the *inverse* map displayed
below, with domain the (x, y)-plane, and target the (X, Y)-plane

$$\phi^{-1}(x, y) = (X, Y). \tag{14.2}$$

The next step is to recall from linear algebra the idea of the *rotation matrix*
through an angle θ, namely the invertible 2×2 matrix

$$R(\theta) = \begin{pmatrix} \cos\theta & -\sin\theta \\ \sin\theta & \cos\theta \end{pmatrix}. \tag{14.3}$$

Note that the rotation matrix corresponding to a zero rotation angle is the identity matrix, which we denote I. We will need to be aware of the additive property of rotation matrices, namely that the rotation matrices through angles θ_1, θ_2 satisfy the following matrix multiplication relations

$$R(\theta_1)R(\theta_2) = R(\theta_1 + \theta_2) = R(\theta_2)R(\theta_1). \tag{14.4}$$

The reader is invited to check these relations, using the sum and difference formulas of elementary trigonometry. In the particular case when the sum of the angles is zero that reduces to the following relations, saying that the inverse of the rotation matrix through an angle θ is the rotation matrix through an angle $-\theta$

$$R(\theta)R(-\theta) = I = R(-\theta)R(\theta). \tag{14.5}$$

The central concept of this chapter is the special type of planar mapping known as a *congruence*, a planar mapping ϕ given by a formula of the following form, where R is a rotation matrix, and T is a fixed vector

$$\phi(Z) = ZR + T. \tag{14.6}$$

We will refer to R and T as the *rotational* and *translational* parts of the congruence. We can express ϕ in the coordinate form (14.1) by setting $T = (u, v)$ and observing that then

$$\begin{cases} x = X \cos\theta - Y \sin\theta + u \\ y = X \sin\theta + Y \cos\theta + v. \end{cases} \tag{14.7}$$

The congruence (14.6) is invertible since the equation $z = \phi(Z)$ can always be solved for Z in terms of z. Doing this explicitly we see that the inverse is another congruence

$$\phi^{-1}(z) = zR^{-1} - TR^{-1}.$$

The inverse congruence can likewise be expressed in the coordinate form (14.1)

$$\begin{cases} X = (x - u) \cos\theta + (y - v) \sin\theta \\ Y = -(x - u) \sin\theta + (y - v) \cos\theta. \end{cases} \tag{14.8}$$

The next two examples represent special cases of general congruences, namely translations and rotations.

Example 14.1 A *translation* through a fixed vector T is the planar mapping ϕ given by $\phi(Z) = Z + T$. Thus translations are the special case of congruences with rotation matrix the identity. One thinks of a translation as a 'sliding' of the whole plane in the direction T. (Figure 14.2 illustrates the effect of a translation

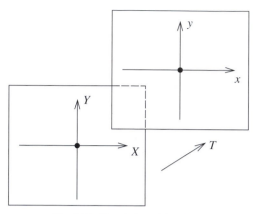

Fig. 14.2. Translation of the plane

on several points.) To write translations and their inverses in the coordinate forms, set $Z = (X, Y)$, $T = (u, v)$ to obtain the formulas

$$x = X + u, \quad y = Y + v; \qquad X = x - u, \quad Y = y - v.$$

Example 14.2 By a *rotation* (about the origin) through an angle θ we mean the planar mapping ϕ defined by $\phi(Z) = ZR$ with R a rotation matrix. Thus rotations are the special case of congruences in which the translational part T is zero. It is worth pointing out that they appear in linear algebra as examples of linear mappings. In coordinate terms, rotations and their inverses can be written in the shape

$$\begin{cases} x = X \cos \theta - Y \sin \theta \\ y = X \sin \theta + Y \cos \theta. \end{cases} \qquad \begin{cases} X = x \cos \theta + y \sin \theta \\ Y = -x \sin \theta + y \cos \theta. \end{cases}$$

To make the abstract idea more concrete, here are some examples of rotations (through special angles) which we will have occasion to use later.

Example 14.3 Rotation about the origin through a right angle, and its inverse are the maps defined by

$$x = -Y, \quad y = X; \qquad X = y, \quad Y = -x.$$

Likewise, rotation about the origin through two right angles, and its inverse are the maps

$$x = -X, \quad y = -Y; \qquad X = -x, \quad Y = -y.$$

Finally, rotation about the origin through $\pi/4$, and its inverse are the mappings

$$x = \frac{X - Y}{\sqrt{2}}, \quad y = \frac{X + Y}{\sqrt{2}}; \qquad X = \frac{x + y}{\sqrt{2}}, \quad Y = \frac{-x + y}{\sqrt{2}}.$$

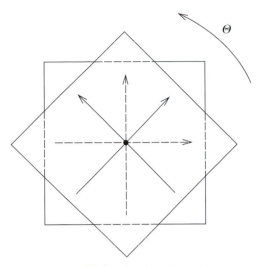

Fig. 14.3. Rotation about the origin

Here is an example of a general congruence which is neither a translation nor a rotation. We first met this example in Chapter 5 when defining the concept of 'centre'.

Example 14.4 Central reflexion in the point (u, v) is the planar map defined by $x = 2u - X$, $y = 2v - Y$. Central reflexion is thus a congruence with rotational angle $\theta = \pi$, and translation through $(2u, 2v)$. Of course when (u, v) is the zero vector, central reflexion is a pure rotation.

An important aspect of congruences is that they form a 'group' in the sense of abstract algebra. It is only the terminology we require, not the theory.

Theorem 14.1 *The congruences form a group, in the following sense. The identity mapping is a congruence, the compositie of two congruences is a congruence, and the inverse of a congruence is a congruence.*

Proof The identity map of the plane is the special case of the general congruence (14.6) obtained by taking the rotation angle to be zero, and the translation vector to be the zero vector. By calculation, the composite of two congruences $\phi_1(Z) = ZR_1 + T_1$, $\phi_2(Z) = ZR_2 + T_2$ is the congruence $\phi(Z) = ZR + T$ where $R = R_2R_1$ and $T = T_2R_1 + T_1$. And, as we observed above, the inverse of a congruence is another congruence. $\square$

Having established some algebraic properties of congruences we need to say something about their metric properties. A key property of a rotation ρ

about the origin is that it leaves scalar products invariant, in the following sense.

Lemma 14.2 *Let ρ be a rotation. Then for any vectors Z, W we have the relation*

$$\rho(Z) \bullet \rho(W) = Z \bullet W. \tag{14.9}$$

Proof The rotation ρ is defined by a formula $\rho(Z) = ZR$ with R a rotation matrix. Write $Z = (x, y)$, $W = (u, v)$. Then, writing $c = \cos\theta$, $s = \sin\theta$ for concision of expression, we have the following relations, and a one-line calculation verifies (14.9)

$$\rho(Z) = (xc - ys, xs + yc), \quad \rho(W) = (uc - vs, us + vc).$$

$\square$

Example 14.5 An immediate consequence is that rotations preserve distance, in the sense that for any vector Z we have

$$|\rho(Z)| = |Z|. \tag{14.10}$$

That follows immdediately from the calculation below on taking positive square roots

$$|\rho(Z)|^2 = \rho(Z) \bullet \rho(Z) = Z \bullet Z = |Z|^2.$$

That leads us to an important general property of congruences. An invertible planar mapping ϕ is an *isometry* when it preserves distance, in the sense that for any points Z, Z' we have

$$|\phi(Z) - \phi(Z')| = |Z - Z'|.$$

Example 14.6 Any congruence ϕ is an isometry. To this end write $\phi(Z) = \rho(Z) + T$ with ρ a rotation, and T a fixed vector. Then, using the facts that rotations are linear and preserve distance, we have

$$|\phi(Z) - \phi(Z')| = |\rho(Z) - \rho(Z')| = |\rho(Z - Z')| = |Z - Z'|.$$

14.2 Congruent Lines

The next step is to be clear about the effect of congruences on lines. Two linear functions L, M are *congruent* when there exists a congruence ϕ and a non-zero constant μ for which $M = \mu(L \circ \phi)$. More explicitly, write $\phi(X, Y) = (x, y)$

so that

$$M(X, Y) = \mu L(x, y) = \mu L(\phi(X, Y)). \qquad (14.11)$$

The reader is left to verify that congruence is an equivalence relation on linear functions. (Exercise 14.2.1.) The proof is a consequence of the fact that the congruences form a group. Moreover, a linear function L is congruent to any constant multiple of L, so the concept of congruence is well defined (and an equivalence relation) on lines. To make the definition more concrete, write $L = ax + by + c$. Then, substituting the expressions for x, y in the general congruence (14.7) we see that we can write $M = AX + BY + C$ where

$$\begin{cases} A = \mu(a\cos\theta + b\sin\theta) \\ B = \mu(-a\sin\theta + b\cos\theta) \\ C = \mu(au + bv + c). \end{cases} \qquad (14.12)$$

Here is an example which takes advantage of these explicit formulas to show that up to congruence, all lines are the same.

Example 14.7 Any two lines L, M are congruent. By transitivity of the relation, it suffices to show that the line L defined by $x = 0$ is congruent to any given line M. Taking $a = 1$, $b = 0$, $c = 0$ in (14.12) we see that $M = AX + BY + C$ where

$$A = \mu\cos\theta, \qquad B = -\mu\sin\theta, \qquad C = \mu u.$$

Given A, B, C (with at least one of A, B non-zero) choose $\mu > 0$ with $\mu^2 = A^2 + B^2$, and then observe that these relations are satisfied for a unique angle θ with $0 \le \theta < 2\pi$, and a unique constant u.

One consequence of (14.11) is that the congruence ϕ maps the zero set of M bijectively to the zero set of L. The next example uses this observation to clarify the effect of a congruence on pencils of lines.

Example 14.8 Let ϕ be a congruence mapping the lines L, M to the lines L', M'. Then a point P is an intersection of L, M if and only if its image P' under ϕ is an intersection of L', M'. Thus L, M intersect if and only if L', M' intersect: likewise L, M are parallel if and only if L', M' are parallel. More generally, ϕ maps the pencil of lines through a point P to the pencil of lines through the image P'; and any parallel pencil of lines to another parallel pencil.

Example 14.9 Congruences preserve midpoints of line segments. Let p, q, r be collinear points mapped by a congruence ϕ to collinear points p', q', r'. Then r is the midpoint of the line segment p, q if and only if r' is the midpoint of the line segment p', q'. We have only to observe that r is equidistant from p, q if and only if r' is equidistant from p', q'.

Example 14.10 Congruences preserve angles between two lines. Let $M = AX + BY + C$, $M' = A'X + B'Y + C'$ be lines, obtained from the lines $L = ax + by + c$, $L' = a'x + b'y + c'$ by a congruence ϕ. Let ρ be the rotational part of ϕ. Then the formulas (14.12) show that the vectors $V = (a, b)$, $V' = (a', b')$ are expressed in terms of the vectors $W = (A, B)$, $W' = (A', B')$ by the relations $V = \mu\rho(W)$, $V' = \mu\rho(W')$ for some non-zero constant μ. The angles θ between the lines L, L' are then determined by

$$\pm\cos\theta = \frac{V \bullet V'}{|V||V'|} = \frac{\rho(W) \bullet \rho(W')}{|\rho(W)||\rho(W')|} = \frac{W \bullet W'}{|W||W'|}.$$

The last equality uses the relations (14.9), (14.10). Thus the angles θ are precisely the angles between the lines M, M'. In particular, congruences preserve perpendicularity between two lines.

Exercises

14.2.1　Show that the relation of congruence is an equivalence relation on lines.

14.2.2　Let P be a point, let L be a line, and let ϕ be a congruence mapping P to a point P' and L to a line L'. Show that the distance from P to L equals the distance from P' to L'.

14.3 Congruent Conics

The definition of 'congruence' in the previous section can be extended from lines to conics. Two quadratic functions Q, R are *congruent* when there exists a congruence ϕ, and a non-zero constant μ, for which $R = \mu(Q \circ \phi)$. More concretely, writing $\phi(X, Y) = (x, y)$ the relation is that

$$R(X, Y) = \mu Q(x, y) = \mu Q(\phi(X, Y)). \tag{14.13}$$

We may on occasion use a finer terminology, for instance that Q, R are *translationally* congruent when ϕ is a translation, or *rotationally* congruent when ϕ is a rotation. In fact we met translational congruence in Chapter 5 when discussing centres, though the idea of rotational congruence is new.

Example 14.11 Consider rotation about the origin through an angle θ defined by $\tan\theta = -4/3$: thus $\sin\theta = -4/5$, $\cos\theta = 3/5$, and the rotation is defined by the formulas

$$x = \frac{3X + 4Y}{5}, \qquad y = \frac{-4X + 3Y}{5}.$$

The result of this rotation on the conic Q below is obtained by substituting for x, y in Q, yielding the rotationally congruent conic R

$$Q(x, y) = 11x^2 + 24xy + 4y^2 - 5, \quad R(X, Y) = -5(X^2 - 4Y^2 + 1).$$

The reader is left to verify that congruence is an equivalence relation on quadratic functions. (Exercise 14.3.1.) The proof is a direct consequence of the fact that the congruences form a group. Moreover, a quadratic function Q is congruent to any constant multiple of Q, so the concept of congruence is well defined (and an equivalence relation) on conics. Congruence is the formal mathematical concept expressing the idea illustrated by Figure 14.1.

Example 14.12 Congruences preserve zero sets of conics, in the following sense. Let Q, R be congruent conics, so there exists a congruence ϕ and a non-zero constant μ for which (14.7) holds. Then the point (x, y) lies in the zero set of R if and only if the point $\phi(x, y)$ lies in the zero set of Q. Put another way, the congruence ϕ maps the zero set of R bijectively to that of Q. In particular the zero set of Q is infinite, a point, or empty if and only if the zero set of R is infinite, a point, or empty. For instance, a real ellipse cannot be congruent to a virtual ellipse, since the zero set of the latter is empty, whilst that of the former is not.

The next example shows that congruences preserve midpoint loci of conics, and hence also their axes and their vertices.

Example 14.13 Consider a conic Q and a parallel pencil of lines M. A congruence ϕ will map Q to a congruent conic Q', and the lines M to a parallel pencil of lines M'. By Example 14.9 the point P is the midpoint of the chord M if and only if its image P' under ϕ is the midpoint of the chord M'. Thus a line L is the midpoint locus for Q (in the direction of the lines M) if and only if the congruent line L' is the midpoint locus for Q' (in the direction of the lines M'). (Figure 14.4.) In fact more is true. By Example 14.10 the lines L, M are perpendicular if and only if L', M' are perpendicular. It follows that L is an axis for Q if and only if L' is an axis for Q'.

The principal objective of the next chapter will be to classify conics up to the relation of congruence. To any conic Q we will associate a 'normal form', a congruent conic given by a simple formula. Quite apart from its intrinsic

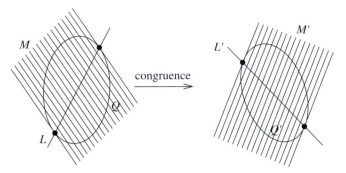

Fig. 14.4. Invariance of midpoint loci

interest, the classification has an important practical merit: the geometry of normal forms can be established via straightforward explicit calculations, and we can then deduce the geometry of arbitrary conics.

Exercises

14.3.1   Show that the relation of congruence is an equivalence relation on conics.

14.3.2   Show that congruences preserve the property of 'reducibility' in the sense that if Q, Q' are congruent conics, then Q is reducible if and only if Q' is reducible.

14.3.3   Show that congruences preserve centres, in the following sense. Let P be a point, let Q be a conic, and let ϕ be a congruence, taking P to a point P', and Q to a conic Q'. Then P is a centre for Q if and only if P' is a centre for Q'.

14.3.4   Show that congruences preserve tangency in the following sense. Let L be a line intersecting a conic Q at a point P, and let ϕ be a congruence taking the line L to a L', the conic Q to Q' and the point P to P'. Then L is the tangent to Q at P if and only if L' is the tangent to Q' at P'.

14.4 The Invariance Theorem

In the next chapter we will classify conics, up to the relation of congruence. That raises natural questions. For instance, do the resulting classes overlap? As we saw above, the zero set provides a crude way of distinguishing some classes from others. However, in this respect the invariants of Section 4.3 are much better tools, the key fact being the Invariance Theorem below. To state the result we need a definition. Two quadratic functions Q, R are *strictly congruent* when they satisfy the relation (14.13) with $\mu = 1$. In more detail, there exists a congruence ϕ for which $R = Q \circ \phi$: thus, if $\phi(X, Y) = (x, y)$ then

$$R(X, Y) = Q(x, y) = Q(\phi(X, Y)). \tag{14.14}$$

Strictly congruent quadratic functions are automatically congruent. However the converse fails. Two quadratic functions may define the same conic, but fail to be strictly congruent. (Exercise 14.4.1.)

Theorem 14.3 *Let Q, Q' be strictly congruent quadratic functions with invariants τ, δ, Δ and τ', δ', Δ'. Then $\tau = \tau'$, $\delta = \delta'$, $\Delta = \Delta'$.* (The Invariance Theorem.)

We will postpone the proof of the Invariance Theorem to Chapter 17. For the moment it is the statement which is important to us, rather than the proof.

Example 14.14 The conics Q, Q' below were obtained from each other in Example 14.11 by applying a rotation

$$Q(x, y) = 11x^2 + 24xy + 4y^2 - 5, \quad Q'(X, Y) = -5(X^2 - 4Y^2 + 1).$$

A moments calculation verifies that the invariants take the following values, confirming the Invariance Theorem for this example

$$\tau = \tau' = 15, \qquad \delta = \delta' = -100, \qquad \Delta = \Delta' = 500.$$

Example 14.15 Strictly congruent quadratic functions Q, Q' have the same eigenvalues. The eigenvalues are the roots of the characteristic equations. By (7.5) these can be written in the form below, where τ, δ are the invariants for Q, and τ', δ' those for Q'. By the Invariance Theorem we have $\tau' = \tau$, $\delta' = \delta$, so the characteristic equations, and hence the eigenvalues, are equal

$$\lambda^2 - \tau\lambda + \delta = 0, \qquad \lambda^2 - \tau'\lambda + \delta' = 0.$$

It is the following consequence of the Invariance Theorem which we use in practice. Let Q, Q' be quadratic functions, with invariants τ, δ, Δ and τ', δ', Δ'. Suppose that the conics they define are congruent, so there exists a congruence ϕ, and a non-zero scalar μ for which

$$Q'(X, Y) = \mu Q(\phi(X, Y)) = \mu Q''(X, Y).$$

By the Invariance Theorem, Q, Q'' have the same invariants. However, Q' is obtained from Q'' on multiplying by μ. Thus we obtain a general principle, that the invariants of congruent conics Q, Q' satisfy relations of the form

$$\tau' = \mu\tau, \qquad \delta' = \mu^2\delta, \qquad \Delta' = \mu^3\Delta. \qquad (14.15)$$

These relations allow us to draw a number of useful conclusions. The discriminants Δ, Δ' of congruent conics Q, Q' are both zero, or both non-zero. Thus Q is degenerate if and only if Q' is degenerate. Moreover, the delta invariants δ, δ' are both zero, or both non-zero; indeed, since μ^2 is positive, they even have the same signs. It follows from these comments that the concepts of ellipse, parabola, and hyperbola defined for non-degenerate conics by Table 4.1 are all invariant under congruence, as are the concepts of real, parallel, and virtual line-pairs defined for degenerate conics in Table 6.1.

Exercises

14.4.1   Let Q, Q' be quadratic functions with $Q' = \mu Q$ for some constant $\mu \neq 0, 1$. Show that Q, Q' cannot be congruent.

14.4.2   Let ϕ be a congruence taking the ellipse Q to another ellipse Q'. Show that ϕ takes the major (respectively minor) axis of Q to the major (respectively minor) axis of Q'.

15

Classifying Conics

The objective of this chapter is to obtain a complete list of conics (up to the relation of congruence) and explore the consequences of that listing. We start with an arbitrary conic Q and reduce the number of terms in it till we reach a 'normal form', a congruent conic given by a particularly simple formula. Section 15.1 is a key step in this process, rotating the axes till they are parallel to the coordinate axes. There are three main cases: Q has a unique centre, a line of centres, or no centre. In each case we will derive a list of 'normal forms'. The net result is a listing into nine 'classes', each of which (with one exception) involves moduli. The question of distinguishing these classes, and ensuring there are no redundancies amongst the normal forms, provides the material for the next chapter.

The classification yields significant gains. The simplicity of the 'normal form' allows us to elucidate its geometry with relative ease. However, the relation of 'congruence' preserves all the desirable geometric features of a conic. Thus in principle we can access the geometry of *any* conic, without recourse to complex calculations. The first gain is the fact that any parabola, real ellipse, or hyperbola is constructible, so the interesting metric geometry of constructible conics can be extended to the three most important conic classes. Another geometric gain is that we can relate eigenvalues to axis lengths, leading to a practical technique for calculating axis lengths directly from equations.

15.1 Rotating the Axes

Consider the general conic $(\star)$ below. The initial step in listing conics up to congruence is based on an observation made in Example 7.9, namely that $(\star)$ has an axis parallel to a coordinate axis if and only if the coefficient $2h$ of the *cross term* xy is zero. That is the motivation for our first result, that by rotation we can force the axes of a conic to be parallel to the coordinate axes

$$Q(x, y) = ax^2 + 2hxy + by^2 + 2gx + 2fy + c. \qquad (\star)$$

Lemma 15.1 *Any conic* $(\star)$ *is rotationally congruent to one in which the coefficient of the cross term xy is zero.*

Proof Consider the effect on $(\star)$ of a rotation through an angle θ, given by the formulas

$$\begin{cases} x = X \cos\theta - Y \sin\theta \\ y = X \sin\theta + Y \cos\theta. \end{cases}$$

Substituting these expressions for x, y in Q, we obtain another conic Q'. Direct calculation (using the double angle formulas of school trigonometry) shows that the coefficient $2h'$ of xy in Q' is given by

$$h' = \frac{1}{2}(b - a)\sin 2\theta + h\cos 2\theta.$$

We seek an angle θ for which $h' = 0$. We can assume $h \neq 0$, else there is nothing to prove. Thus when $a = b$ the condition $h' = 0$ reduces to $\cos 2\theta = 0$, and we can take $\theta = \pi/4$: otherwise, a suitable angle θ is determined by

$$\tan 2\theta = \frac{2h}{a - b}.$$

$\square$

To write down the rotation explicitly in an example requires the values of $\sin\theta$ and $\cos\theta$ which are easily deduced from $\tan\theta$. However, the proof produces a value of $\tan 2\theta$. The technique is to use the trigonometric identity

$$\tan 2\theta = \frac{2\tan\theta}{1 - \tan^2\theta}.$$

That produces a quadratic in $T = \tan\theta$ which can be solved explicitly in a given example, namely $hT^2 + (a - b)T - h = 0$. It has a positive discriminant, so there are two distinct real roots. The mechanics of the calculation are illustrated in the next example.

Example 15.1 For the conic Q below the required angle θ is determined by the relation $\tan 2\theta = -24/7$

$$Q(x, y) = 41x^2 - 24xy + 34y^2 - 90x + 5y + 25.$$

Setting $T = \tan\theta$ we obtain the quadratic $12T^2 - 7T - 12 = 0$ with roots

$T = 4/3$, $T = -3/4$. When $\tan\theta = 4/3$ we have $\sin\theta = 4/5$, $\cos\theta = 3/5$ and the corresponding rotation is

$$x = \frac{3X - 4Y}{5}, \qquad y = \frac{-4X + 3Y}{5}.$$

Substituting for x, y in $Q(x, y)$ we obtain the following 'rotated' form, in which the cross term is absent

$$Q'(X, Y) = X^2 + 2Y^2 - 2X + 3Y + 1.$$

Exercises

15.1.1   In each of the following cases find a rotation forcing the cross term in Q to vanish:

   (i) $Q = 2x^2 - 3xy + 2y^2 - 1$,
   (ii) $Q = 6x^2 + 24xy - y^2$,
   (iii) $Q = 11x^2 + 4xy + 14y^2 - 5$.

15.1.2   In each of the following cases find a rotation forcing the cross term in Q to vanish:

   (i) $Q = 9x^2 + 24xy + 2y^2 - 6x + 20y + 41$,
   (ii) $Q = 8x^2 + 12xy - 8y^2 + 12x + 4y + 3$,
   (iii) $Q = 9x^2 - 4xy + 6y^2 - 10x - 7$.

15.2 Listing Normal Forms

The net result of the previous section is that any conic is congruent to one in which the cross term is absent, so having the shape

$$Ax^2 + By^2 + 2Gx + 2Fy + C \qquad\qquad (\star\star)$$

with at least one of A, B non-zero. The geometry of this conic depends crucially on whether it has a unique centre, a line of centres, or no centre. It will help to recall at this juncture that by Lemma 5.3 the centres are the solutions of the following equations

$$Ax + G = 0, \quad By + F = 0. \qquad\qquad (15.1)$$

Theorem 15.2 *Any non-degenerate conic* $(\star)$ *with a unique centre is congruent to one of the normal forms below, where a, b are positive constants*

$$\frac{x^2}{a^2} + \frac{y^2}{b^2} = 1, \qquad \frac{x^2}{a^2} + \frac{y^2}{b^2} = -1, \qquad \frac{x^2}{a^2} - \frac{y^2}{b^2} = 1.$$

Proof Translating the centre of (⋆) to the origin, and then rotating to get rid of the cross term we obtain a conic of the following form, for some constant K

$$Ax^2 + By^2 - K.$$

Note first that A, B are both non-zero, else the conic fails to have a unique centre. We can suppose A is positive (multiplying through by -1 if necessary) allowing us to define a constant a by the relation $Aa^2 = 1$. Moreover, K is non-zero, since the discriminant is $-ABK$ and the conic is assumed to be non-degenerate. Further, we can suppose $K = 1$ when K is positive (dividing through by K) or $K = -1$ when K is negative (dividing through by $-K$). When B is positive, define a constant $b > 0$ by $Bb^2 = 1$. That gives us the first two normal forms. When B is negative, define b by $Bb^2 = -1$. In the case $K = 1$ that produces the third normal form. The case $K = -1$ is congruent to the third normal form: just multiply through by -1 and rotate through a right angle (changing the variables x, y to $-y$, x) to get back to the case $K = 1$.
□

The first two normal forms of Theorem 15.2 represent the classes of real and virtual ellipses, whilst the third represents the class of hyperbolas. There is detail here worth spelling out.

Lemma 15.3 *Any real circle, virtual circle, or rectangular hyperbola is congruent to one of the following normal forms, where a is a positive constant*

$$x^2 + y^2 = a^2, \qquad x^2 + y^2 = -a^2, \qquad x^2 - y^2 = a^2.$$

Proof The real circle, virtual circle, and rectangular hyperbola types are all invariant under congruence. Moreover, they are all non-degenerate types with a unique centre, so congruent to one of the normal forms listed in Theorem 15.2. It remains to observe that in each case the normal form is of the given type if and only if $a = b$.
□

Lemma 15.4 *Any non-circular real or virtual ellipse is congruent to one of the following normal forms*

$$\frac{x^2}{a^2} + \frac{y^2}{b^2} = 1, \qquad \frac{x^2}{a^2} + \frac{y^2}{b^2} = -1, \qquad (0 < b < a).$$

Proof Both types in the statement are invariant under congruence, and congruent to the first two families of normal forms in Theorem 15.2 with distinct moduli a, b. Further, we can assume $0 < b < a$: indeed we need only rotate the conic through a right angle to obtain a normal form in which the the roles of a, b are switched.   □

Of course, the first normal forms are the standard real ellipses first introduced in Section 4.1. It would be appropriate to dub the second normal forms the *standard* virtual ellipses. The reader is warned that the arguments for ellipses do not apply to hyperbolas. There is no reason to suppose that switching the moduli a, b in a normal form for a hyperbola will produce a congruent conic. In fact, as we will see in the next chapter, that is not the case.

Theorem 15.5   *Any parabola* (⋆) *is congruent to a standard parabola* $y^2 = 4ax$ *with a* > 0.

Proof   In Example 5.6 we observed that parabolas are automatically non-central conics. The condition for (⋆⋆) to be non-central is that the equations (15.1) have no solution. That is the case when either $A \neq 0$, $B = 0$, $F \neq 0$ or $A = 0$, $B \neq 0$, $G \neq 0$. The former possibility reduces to the latter on replacing x, y by $-y$, x (a rotation through a right angle), so we need only consider the latter. Dividing through by B we can suppose that $B = 1$. Translation through $(0, -F)$ then forces the coefficient of y to vanish, resulting in a conic of the form $y^2 + 2Gx + K$ for some new constant K; and then a translation parallel to the x-axis forces that constant to vanish, producing a form $y^2 + 2Gx$. It is no restriction to suppose that $G < 0$: otherwise, we replace x, y by $-x$, $-y$ (a rotation through two right angles) to change the sign of G. Finally, setting $G = -2a$ with $a > 0$ we obtain the normal form of the statement.   □

The above results yield a complete classification of non-degenerate conics up to congruence. It is time to turn our attention to the degenerate cases. First, we deal with degenerate conics having a unique centre.

Theorem 15.6   *Any degenerate conic* (⋆) *with a unique centre is congruent to a real line-pair* $y^2 = c^2 x^2$ *or a virtual line-pair* $y^2 = -c^2 x^2$ *with* $0 < c \leq 1$.

Proof   The initial steps are identical to those in the proof of Theorem 15.2. By translation and rotation we can assume the conic has the following form, for some constant K

$$Ax^2 + By^2 - K.$$

Then A, B are non-zero, else the conic fails to have a unique centre. The discriminant is ABK, so the condition for the conic to be degenerate is that $K = 0$. Dividing the equation through by B we obtain an equation $Cx^2 + y^2$

with C non-zero. When $C < 0$ we set $C = -c^2$, and when $C > 0$ we set $C = c^2$, to obtain the normal forms of the statement. If $c \ge 1$ we apply a rotation through a right angle (switching the roles of the variables) to obtain a normal form with $0 < c \le 1$. $\square$

To complete the classification of conics it remains to discuss conics having a line of centres.

Theorem 15.7 *Any conic* (⋆) *having a line of centres is congruent to the real parallel lines* $y^2 = k^2$, *the virtual parallel lines* $y^2 = -k^2$, *or the repeated line* $y^2 = 0$, *where k is a positive constant.*

Proof In view of the equations (15.1) we have a line of centres if and only if $A = 0$, $B \ne 0$, $G = 0$, or $A \ne 0$, $B = 0$, $F = 0$. The latter possibility reduces to the former on replacing x, y by $-y$, x (a rotation through a right angle) so we need only consider the former. Translating one of the centres to the origin we can assume the conic has the form $Bx^2 + K$ for some constant K. Dividing through by B we can suppose $B = 1$. When $K < 0$ we set $K = -k^2$ with $k > 0$ to obtain a real line-pair: when $K = 0$ we obtain the repeated line; and when $K > 0$ we set $K = k^2$ with $k > 0$, to obtain a virtual line-pair. $\square$

Exercises

15.2.1   Show that $Q = 9x^2 + 24xy + 2y^2 - 6x + 20y + 41$ has a unique centre, and find the equation of the conic obtained by translating the centre to the origin. Find an explicit rotation removing the cross term, and hence determine a normal form for Q.

15.2.2   Let Q be a degenerate conic. Show that the trace invariant τ vanishes if and only if Q is a real line-pair with perpendicular components.

15.2.3   Show that any conic $Q = \alpha L^2 + \beta M^2$, with L, M lines and α, β constants, is a (real or virtual) line-pair, a (real or virtual) parallel line-pair, or a repeated line.

15.3 Some Consequences

The principal gain in listing conics is that the geometry of an arbitrary conic is reduced to that of a normal form. And the simplicity of normal forms means that calculations with them are much easier to carry out. It is of course the real ellipses, parabolas, and hyperbolas which are of greatest interest in this respect. Our first result is that congruences preserve the notion of 'constructibility'.

Lemma 15.8 *Let Q be a constructible conic with eccentricity e. Then any conic Q′ congruent to Q is also constructible, with the same eccentricity e.*

Proof The assumption is that Q arises from the construction comprising a focus F, a directrix D, and eccentricity e. Let ϕ be the congruence taking Q to Q'. Then ϕ maps F to a point F', and D to a line D' not passing through F'. Now let P be a point on Q mapped by ϕ to a point P' on Q'. Since congruences preserve distance, we have $PF = P'F'$, $PD = P'D'$. Then (8.1) yields the relation $P'F'^2 = e^2 P'D'^2$, ensuring that Q' arises from the construction comprising the focus F', the directrix D', and the same eccentricity e. $\square$

The main consequence of this result is that any real ellipse, parabola, or hyperbola is a constructible conic. By Theorem 15.2 such conics are congruent to the standard conics, and by the results of Chapter 8 all the standard conics are constructible. Indeed the argument establishes more: any parabola has a unique construction, whilst any real ellipse or hyperbola has precisely two constructions.

Example 15.2 In Lemma 8.4 we showed that any standard real ellipse has just two constructions, each giving rise to a focus on the major axis, and a directrix perpendicular to that axis. Since the concepts of focus, directrix, and major axis are all invariant under congruences, the same is true of any real ellipse.

In principle any of the focal properties established for standard conics, also hold for arbitrary conics. For instance all three conic classes share the reflective properties established for the standard conics. Likewise, all ellipses have the property that the sum of the distances from a general point to the foci is constant, and equal to the length of the major axis.

Exercises

15.3.1   Verify that the zero set of any normal form is infinite, a single point, or empty and deduce that the zero set of *any* conic is infinite, a single point, or empty.

15.3.2   Two congruent parabolas Q, Q' have a common focus. Show that their common chord passes through the focus, and is a bisector of the axes.

15.4 Eigenvalues and Axes

Eigenvalues and eigenvectors arose naturally in Chapter 7 when studying axes of conics. Recall that the eigenvectors give the directions of the axes. However, the significance of the eigenvalues is less clear. The purpose of this section is to clarify the situation by relating the eigenvalues to axis lengths. Quite apart from the conceptual gain, that has the practical advantage of enabling us to calculate axis lengths directly from an equation, without having to determine a normal form.

Let us start by spelling out the effects of a congruence on eigenvalues. According to Example 14.15 strictly congruent quadratic functions have the same eigenvalues. However, eigenvalues are not invariant under general congruences. Let Q be a quadratic function Q with invariants τ, δ, Δ. As we pointed out in Chapter 7, multiplication of Q by a constant k multiplies any eigenvalue by k. The general situation is this. Any congruent quadratic function Q' has the form $Q' = kQ''$ with Q'' strictly congruent to Q. Now Q, Q'' have the same invariants (by the Invariance Theorem) so those of Q' are $k\tau$, $k\delta$, $k\Delta$. Moreover, any eigenvalue λ for Q corresponds to a unique eigenvalue $k\lambda$ for Q'.

Against this background, consider conics Q having a unique centre. (Thus their delta invariants and eigenvalues are non-zero.) The classification produces a normal form with centre the origin, and constant term Δ/δ. (Lemma 5.5.) Of course, the constant term changes under a congruence. Now let λ be an eigenvalue of Q. In view of the above discussion the expression $\Delta/\lambda\delta$ is the same for the normal form, provided we choose its eigenvalue corresponding to λ. For real ellipses that has the following consequence.

Lemma 15.9 *Let λ be an eigenvalue of a real ellipse Q. Then the semilength s of the corresponding axis is given by*

$$s^2 = \left| \frac{\Delta}{\delta\lambda} \right|. \tag{15.2}$$

Proof In view of the above remarks both sides of the displayed formula are invariant under congruences, so it suffices to establish the formula for the standard real ellipse with moduli a, b satisfying $0 < b < a$. For that form the semilengths of the major and minor axes are $s = a, b$. Moreover setting $A = 1/a^2$, $B = 1/b^2$ we find that the invariants are $\delta = AB$, $\Delta = -AB$ and the eigenvalues are $\lambda = A, B$. Thus the RHS of the formula has the value a^2 for $\lambda = A$, and b^2 for $\lambda = B$, establishing the result. $\qquad\square$

This result confirms an earlier statement, namely that the major axis of an ellipse corresponds to the eigenvalue of smaller absolute value, whilst the minor axis corresponds to the eigenvalue of greater absolute value.

Example 15.3 The invariants of the ellipse Q defined below are readily checked to be $\tau = 6$, $\delta = 8$, $\Delta = -1024$

$$Q = 3x^2 + 2xy + 3y^2 - 6x + 14y - 101.$$

The characteristic equation is $\lambda^2 - 6\lambda + 8 = 0$ producing the positive eigenvalues $\lambda = 2, 4$. According to (15.2), the semilengths s_1, s_2 of the axes are given by $s_1^2 = 32$, $s_2^2 = 64$ yielding $s_1 = 4\sqrt{2}$, $s_2 = 8$.

Likewise, we can apply these ideas to hyperbolas to obtain two useful consequences.

Lemma 15.10 *The transverse (resp. conjugate) axis of a hyperbola Q corresponds to the eigenvalue λ having the opposite (resp. same) sign as Δ/δ, and has semilength s given by (15.2).*

Proof The transverse and conjugate axes are preserved by congruences, so it suffices to check the statements for the standard hyperbolas with moduli a, b. We keep to the notation of the previous proof. The eigenvalues are then $\lambda = A$, $\lambda = -B$ with the positive (resp. negative) eigenvalue corresponding to the transverse (resp. conjugate) axis. The first statement then follows from the fact that $\Delta/\lambda\delta$ is negative (resp. negative) for the positive (resp. negative) eigenvalue. The formula for the semilength follows exactly as in the ellipse case. $\qquad\square$

Example 15.4 The conic $Q = 5x^2 - 24xy - 5y^2 + 14x + 8y - 16$ has invariants $\tau = 0$, $\delta = -169$, $\Delta = 2197$ so is a rectangular hyperbola with $\Delta/\delta = -13$. The characteristic equation is $\lambda^2 - 169 = 0$ so the eigenvalues are $\lambda = 13, -13$. In view of the above result the transverse axis corresponds to the positive eigenvalue $\lambda = 13$. Moreover, according to (15.2) its semilength s is given by $s^2 = 1$, so $s = 1$. The reader is left to verify that the transverse axis is $2x + 3y - 2 = 0$, and that the conjugate axis is $3x - 2y + 1 = 0$.

Exercises

15.4.1 In each of the following cases find the major and minor axes of the given ellipse and their semilengths:

    (i) $5x^2 - 6xy + 5y^2 + 18x - 14y + 9 = 0$,

    (ii) $13x^2 - 32xy + 37y^2 - 14x - 34y - 35 = 0$,

    (iii) $3x^2 + 2xy + 3y^2 + 14x + 20y - 183 = 0$.

15.4.2 In each of the following cases find the transverse and conjugate axes
of the given hyperbola, and the semilength of the transverse axis:

(i) $3x^2 - 10xy + 3y^2 + 16x - 16y + 8 = 0$,
(ii) $x^2 - 6xy - 7y^2 - 16x - 48y - 88 = 0$,
(iii) $4x^2 - 10xy + 4y^2 + 6x - 12y - 9 = 0$.

16

Distinguishing Conics

The net result of the classification in Chapter 15 is a list of nine basic classes of conics, all of which (with the sole exception of the repeated line) involve moduli. The listing raises two natural questions, providing the material for this chapter. The first is whether any of the lists overlap: can a conic be congruent to normal forms in two *different* classes? As we shall see, that cannot happen, but it requires proof. Together with the zero sets, the invariants enable us to distinguish all nine classes. The net result is a simple, efficient recognition technique. Section 16.2 is in the nature of an extended example, illustrating the application of these ideas to the classical Greek construction of conics, as plane sections of a fixed cone. The second question is whether there is duplication within a class: can a conic be congruent to two normal forms within the *same* class? Again, that cannot happen, but it does require proof.

16.1 Distinguishing Classes

The normal forms for the nine main conic classes, their associated invariants and the cardinal of their zero sets, are listed in Table 16.1. Of the four non-degenerate classes, just one (the hyperbola) has $\delta < 0$, just one (the parabola) has $\delta = 0$, whilst two (the real and virtual ellipses) have $\delta > 0$. However, by definition, the real and virtual ellipses are distinguished by their zero sets. Thus no two of the four classes can be congruent. Of the degenerate classes, just one (the real line-pair) has $\delta < 0$, just one (the virtual line-pair) has $\delta > 0$, whilst the remaining three all have $\delta = 0$. The class of virtual parallel lines is distinguished from the other two by the fact that the zero set is empty. Finally, a repeated line cannot be congruent to real parallel lines, since in the former case every point is singular, and in the latter case no point is singular.

Table 16.1. *Invariants for conic classes*

class	Δ	δ	zero set
real ellipses	$\Delta \neq 0$	$\delta > 0$	infinite
virtual ellipses	$\Delta \neq 0$	$\delta > 0$	empty
hyperbolas	$\Delta \neq 0$	$\delta < 0$	infinite
parabolas	$\Delta \neq 0$	$\delta = 0$	infinite
real line-pairs	$\Delta = 0$	$\delta < 0$	infinite
virtual line-pairs	$\Delta = 0$	$\delta > 0$	point
real parallel lines	$\Delta = 0$	$\delta = 0$	infinite
virtual parallel lines	$\Delta = 0$	$\delta = 0$	empty
repeated lines	$\Delta = 0$	$\delta = 0$	infinite

We can now predict the class of a conic largely by calculating its invariants. The next two examples illustrate situations where a knowledge of the invariants alone is not sufficient to determine the class of a conic; however, even a minimal knowledge of the zero set is sufficient to resolve the question.

Example 16.1 For $Q(x, y) = x^2 - 2xy + 5y^2 + 2x - 10y + 1$ the invariants are $\tau = 6$, $\delta = 4$, $\Delta = 24$, so by Table 16.1 the conic is a real or virtual ellipse. In fact it is a real ellipse because its zero set contains a point. One way of seeing that is to intersect it with the line $y = 0$: that gives $x^2 + 2x + 1 = 0$, i.e. $(x + 1)^2 = 0$, yielding $x = -1$. Thus $(-1, 0)$ is a point on Q.

Example 16.2 For the family of conics $Q(x, y) = y^2 - \alpha x^2 - 2x$ we have $\tau = 1 - \alpha$, $\delta = -\alpha$, $\Delta = 1$. Note that Q always passes through the origin, so cannot have an empty zero set, and Table 16.1 determines the class. Indeed Q is a real ellipse for $\alpha < 0$, a parabola for $\alpha = 0$, and a hyperbola for $\alpha > 0$. It is illuminating to see how Q changes as α varies. When $\alpha = -1$ we have a circle: in the range $-1 < \alpha < 0$ we have a real ellipse, which becomes a parabola when $\alpha = 0$; and this changes into a hyperbola as α becomes positive, the branches becoming ever closer as $\alpha \to \infty$.

Exercises

16.1.1   Calculate the invariants for the conic Q below. Show that Q is a parabola if and only if $\alpha = 0$, $\beta \neq 0$, and is a repeated line if and only if $\alpha = 0$, $\beta = 0$, $\gamma = 0$

$$Q(x, y) = y^2 - \alpha x^2 - 2\beta x - \gamma.$$

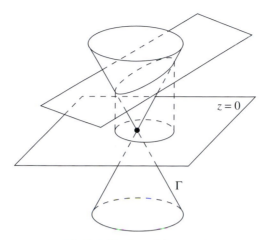

Fig. 16.1. Sections of the cone Γ

16.1.2 In each of the following cases, calculate the invariants of the conic Q_t, and hence determine its class:

(i) $Q_t(x, y) = 2y(x - 1) + 2t(x - y),$
(ii) $Q_t(x, y) = x^2 + t(t + 1)y^2 + 2 - 2txy + 2x,$
(iii) $Q_t(x, y) = t(x^2 + y^2) - (x - 1)^2.$

16.2 Conic Sections

This section is in the nature of an extended example, illustrating how invariants can identify naturally appearing classes of conics. The ancient Greeks envisaged conics as plane sections of a fixed cone. In this section we will show that (with the sole exception of the parallel line-pair) this produces all classes of conics having a non-empty zero set. The cone in question is the set

$$\Gamma = \{(x, y, z) \in \mathbb{R}^3 : x^2 + y^2 = z^2\}.$$

We can visualize Γ by thinking in terms of its sections with a 'horizontal' plane $z = r$, where r is any fixed real number. For $r \neq 0$ the section is the circle $x^2 + y^2 = r^2$, $z = r$ centred at the point $(0, 0, r)$ on the z-axis, and having radius $|r|$; and for $r = 0$ it comprises the origin. (Figure 16.1.)

Now consider the sections of Γ with 'non-vertical' planes having equation $z = \alpha x + \beta y + \gamma$. When we look at such a section from a distant point on the z-axis we see a curve, which we can place in the (x, y)-plane. That curve is obtained by eliminating z between the equations of the plane and the cone,

Table 16.2. *Conic sections*

affine type	Δ	δ
real ellipses	$\gamma \neq 0$	$\alpha^2 + \beta^2 < 1$
hyperbolas	$\gamma \neq 0$	$\alpha^2 + \beta^2 > 1$
parabolas	$\gamma \neq 0$	$\alpha^2 + \beta^2 = 1$
virtual line-pairs	$\gamma = 0$	$\alpha^2 + \beta^2 < 1$
real line-pairs	$\gamma = 0$	$\alpha^2 + \beta^2 > 1$
repeated lines	$\gamma = 0$	$\alpha^2 + \beta^2 = 1$

producing the conic S defined by

$$(\alpha^2 - 1)x^2 + 2\alpha\beta xy + (\beta^2 - 1)y^2 + 2\alpha\gamma x + 2\beta\gamma y + \gamma^2.$$

We will list the possible classes as α, β, γ vary. The reader is left to check that

$$\tau = \alpha^2 + \beta^2 - 2, \qquad \delta = 1 - \alpha^2 - \beta^2, \qquad \Delta = \gamma^2.$$

To determine the class in which S lies, note first that *the zero set of S cannot be empty*: for instance, setting $x = 0$ yields a quadratic in y with discriminant $\gamma^2 \geq 0$, ensuring the existence of a zero. Thus S cannot lie in either of the two conic classes with empty zero sets. The invariants determine the remaining seven classes in Table 16.1 with one exception, namely $\delta = 0$, $\Delta = 0$. In that case $\alpha^2 + \beta^2 = 1$, $\gamma = 0$, so S is a binary quadratic with $\delta = 0$, and therefore a repeated line. The six possibilities are listed in Table 16.2. The condition $\gamma = 0$ is that the plane $z = \alpha x + \beta y + \gamma$ passes through the origin. It then lies outside the cone (virtual line-pair), or cuts the cone (real line-pair), or intersects it along a line (repeated line).

16.3 Conics within a Class

That brings us to the question whether there are congruent normal forms within any of the nine main conic classes. With one exception (repeated lines) all nine classes have normal forms involving moduli, for instance a, b in the classes of real and virtual ellipses. The objective is to show that in a given class congruent normal forms have the same moduli. For some classes simple observations suffice to achieve this objective. It is markedly easier to deal with the cases when the zero set is infinite. The point is best made by looking first at the non-degenerate classes.

Example 16.3 A congruence between two constructible conics will leave the distance between focus and vertex invariant. In particular, that is the case for two congruent parabolas. For a normal form $y^2 = 4ax$ with modulus a that distance is a. It follows that two normal forms are congruent if and only if the moduli are equal.

Example 16.4 The same line of thought can be adopted when dealing with real ellipses. Congruences leave major and minor axes invariant, hence also the distances between vertices on those axes. In the case of a standard real ellipse with moduli a, b satisfying $0 < b < a$ the distance between the major vertices is $2a$, whilst that between the minor vertices is $2b$. Again, we can conclude that two normal forms are congruent if and only if their moduli are equal.

Example 16.5 For hyperbolas we need to be slightly more inventive. Consider two standard hyperbolas with moduli a, b and a', b'. Recall first that the eccentricity is invariant under congruences. The eccentricities e, e' are defined by the relations

$$a^2 e^2 = a^2 - b^2, \qquad a'^2 e'^2 = a'^2 - b'^2.$$

Suppose now that the two hyperbolas are congruent: then we know that $e = e'$. However, congruences also leave invariant the distance between the two vertices on the transverse axis. For the two standard hyperbolas those distances are $2a$, $2a'$ so $a = a'$. It now follows from the displayed formulas that $b^2 = b'^2$ and hence $b = b'$.

However, the above arguments do not apply to virtual ellipses. We could go back to first principles, but it is more productive to consider the invariants τ, δ, Δ. Recall the general principle expressed by (14.2), namely that if Q, Q' are conics with invariants τ, δ, Δ and τ', δ', Δ' there is a non-zero constant κ for which

$$\tau' = \kappa\tau, \qquad \delta' = \kappa^2\delta, \qquad \Delta' = \kappa^3\Delta. \tag{16.1}$$

Lemma 16.1 *Let Q, R be congruent conics, defined by the formulas below, where all the coefficients are non-zero and either $p \le q$, $r \le s$ or $p \ge q$, $r \ge s$: then $p = r$ and $q = s$*

$$Q(x, y) = px^2 + qy^2 + t, \qquad R(x, y) = rx^2 + sy^2 + t.$$

Proof The invariants τ, δ, Δ for Q are $p + q$, pq, pqt, whilst those for R are $r + s$, rs, rst. By (14.2) there is a non-zero scalar λ with

$$r + s = \lambda(p + q), \qquad rs = \lambda^2 pq, \qquad rst = \lambda^3 pqt.$$

The last two relations imply that $\lambda = 1$. Then, elimination of s from the first two yields $(p - r)(q - r) = 0$, so either $p = r$ (and hence $q = s$, as required) or $q = r$ (and hence $p = s$). In the latter case either pair of inequalities yields $p = q = r = s$. $\square$

Example 16.6 Virtual ellipses have normal forms $px^2 + qy^2 + t$, where $p = 1/a^2$, $q = 1/b^2$, $t = 1$ and $0 < b \le a$. Then $p \le q$, and the above result shows that the congruence type determines p, q, and hence the moduli a, b. Exactly the same argument applies to real ellipses when we take $t = -1$.

That brings us to the question of normal forms for degenerate conics. Let us start with a geometrically compelling example, namely real line-pairs. By Example 14.10 the angles between the two component lines are invariant under congruence. We use that fact in the next example, to deduce that congruent normal forms have the same modulus.

Example 16.7 Consider a normal form $y^2 = c^2 x^2$ with $0 < c \le 1$ for a real line-pair obtained in Theorem 15.6. The component lines are $cx - y$, $cx + y$. Using the relation (2.1) we see that the angles θ, ϕ between them are determined by the relations $\cos \theta = \rho(c)$, $\cos \phi = -\rho(c)$ where

$$\rho(c) = \frac{c^2 - 1}{c^2 + 1}.$$

By the above remarks, the angles between the component lines of a congruent normal form $y^2 = d^2 x^2$ with $0 < d \le 1$ are also θ, ϕ, so we have a relation $\rho(c) = \pm\rho(d)$. In either case a short calculation, using the inequalities $0 < c, d \le 1$, leads to the conclusion $c = d$.

However, this approach breaks down for virtual line-pairs. A better way forward is to exploit (16.1). Although τ, δ, Δ are not invariants of *conics*, they give rise to expressions which are. For instance it is clear from (16.1) that the ratio $\omega = \tau^2 : \delta$ is an invariant of conics. (The expressions τ, δ cannot vanish simultaneously.) Here is an alternative derivation of the fact that congruent normal forms for real line-pairs have equal moduli.

Example 16.8 The normal forms $y^2 - c^2x^2$ with $0 < c \le 1$ for real line-pairs have invariants $\tau = 1 - c^2$, $\delta = -c^2$, $\Delta = 0$. Thus the invariant σ for such normal forms is the expression

$$\sigma(c) = -\left(\frac{1 - c^2}{c}\right)^2.$$

Then given two normal forms with moduli c, d we have $\sigma(c) = \sigma(d)$. Using the inequalities $0 < c, d \le 1$ we deduce that $(c - d)(cd + 1) = 0$, and hence that $c = d$.

Example 16.9 The argument of the previous example applies equally well to normal forms $y^2 + c^2x^2$ with $0 < c \le 1$ for virtual line-pairs. In that case $\tau = 1 + c^2$, $\delta = c^2$, $\Delta = 0$ and the corresponding invariant is

$$\sigma(c) = \left(\frac{1 + c^2}{c}\right)^2.$$

However, invariants fail to distinguish parallel line-pairs. For any normal form the invariants are $\tau = 1$, $\delta = 0$, $\Delta = 0$ so are independent of the modulus. For *real* parallel line-pairs there is a very simple way forward. A congruence between two real parallel line-pairs will leave the distance between the lines invariant. For normal forms $y^2 - k^2$ with modulus $k > 0$ that distance is $2k$. Thus two normal forms are congruent if and only if the moduli are equal. However, that argument is not open to us when considering *virtual* line-pairs having normal forms $y^2 + k^2$ with modulus $k > 0$. In such a case we have little choice other than to return to the definitions, and use a little ingenuity.

Example 16.10 Consider two normal forms $y^2 + j^2$ and $y^2 + k^2$ with $j, k > 0$ for virtual line-pairs. We claim that if they are congruent then $j = k$. The result of applying a general congruence (14.7) to $y^2 + k^2$ is to replace y by an expression $sx + cy + v$, where $s = \sin\theta$, $c = \cos\theta$. That produces the conic

$$s^2x^2 + 2scxy + c^2y^2 + 2svx + 2cvy + (v^2 - k^2).$$

For this to be a constant multiple of $y^2 + j^2$ all the coefficients of x^2, xy, x, y must be zero, which is equivalent to the conditions $s = 0$, $v = 0$. Thus we require $y^2 - k^2$ to be a constant multiple of $y^2 - j^2$. Clearly, that can only be the case when $j^2 = k^2$, and hence $j = k$.

Exercises

16.3.1 Show that the conclusion of Lemma 16.1 holds when $t = 0$, provided $p + q = 1, r + s = 1$. Deduce that congruent normal forms for (real or virtual) line-pairs have the same moduli.

16.3.2 Show that the expression Δ/τ^3 is an invariant of conics. Use this to show that congruent normal forms for parabolas have the same moduli.

17

Uniqueness and Invariance

In this final chapter we give proofs of two central results. The first is the Uniqueness Theorem of Chapter 4, that two conics having the same infinite zero set are equal. And the second is the Invariance Theorem of Chapter 16.

17.1 Proof of Uniqueness

The Uniqueness Theorem for conics mimics the model provided by lines, namely that two lines L, L' having the same zero set coincide. However, the corresponding statement for conics Q, Q' is false, as was exemplified by point and virtual circles. The underlying problem with such examples is that the zero sets fail to be infinite. When we restrict ourselves to conics with infinite zero sets the analogous result does hold.

Theorem 17.1 *Let Q, Q' be conics having the same zero set. Then Q, Q' coincide, provided the common zero set is infinite.* (The Uniqueness Theorem.)

Here is the proof for *reducible* conics. It uses no more than the Component Lemma, and the uniqueness result for lines.

Proof Since Q is reducible we can write $Q = LM$ with L, M lines. Every point on L lies on Q, hence on Q'. Then the Component Lemma tells us that L is a line component of Q', so $Q' = LM'$ for some line M'. Suppose first that L, M intersect in a single point, or are parallel. Choose two points on M, not on L: then those points must lie on M'. Since lines are determined by their zero sets (Theorem 1.1) it follows that M, M' are scalar multiples, i.e. $M' = \lambda M$ for some scalar $\lambda \neq 0$: thus $Q' = \lambda LM = \lambda Q$, i.e. Q, Q' are scalar multiples. It remains to consider the case when Q is a repeated line, so L, M are scalar multiples. In that case L, M' must be scalar multiples (so $M' = \mu L$ for some scalar $\mu \neq 0$) else we can find a point on M' not on L, hence a point on Q' not on Q. Then $Q' = \mu L^2 = \mu Q$, so again Q, Q' are scalar multiples. $\qquad\square$

Now consider the *irreducible* case. It is revealing to extend to general conics the same simple-minded approach we used for lines and real circles. The key is a general principle in linear algebra, that N linear conditions on $(N+1)$ unknowns have a non-trivial solution: and, in addition, if the conditions are linearly independent, then that solution is determined up to a scalar multiple. In Theorem 1.1 we applied this fact to two linear conditions on the three coefficients a, b, c in a line $L = ax + by + c$, namely the conditions that L should pass through two distinct points. The principle told us that there is a non-trivial solution; moreover, the conditions were linearly independent, so the solution was determined up to a scalar multiple. The idea is to apply this same line of thinking to a general conic with six coefficients

$$Q(x, y) = ax^2 + 2hxy + by^2 + 2gx + 2fy + c. \qquad (\star)$$

We can find five linear conditions on the coefficients by requiring Q to pass through five points $Z_k = (x_k, y_k)$ with $k = 1, \ldots, 5$

$$ax_1^2 + 2hx_1y_1 + by_1^2 + 2gx_1 + 2fy_1 + c = 0$$

$$\vdots \quad \vdots \quad \vdots \qquad\qquad (17.1)$$

$$ax_5^2 + 2hx_5y_5 + by_5^2 + 2gx_5 + 2fy_5 + c = 0.$$

Our general principle tells us that these equations have a non-trivial solution. (Put another way, through any five points in the plane there passes at least one conic.) The serious question is whether the conditions (17.1) are linearly independent. That brings us to the proof of the Uniqueness Theorem in the irreducible case.

Proof By hypothesis, the zero set of Q is infinite, so contains at least five distinct points Z_1, Z_2, Z_3, Z_4, Z_5. It suffices to show that the linear equations (17.1) are linearly independent, and hence that any conic having the same zero set is equivalent to Q. Suppose otherwise, so that at least one equation is a non-trivial linear combination of the others. By symmetry, we can suppose that the last equation is a non-trivial linear combination of the first four. That means that *any* conic passing through Z_1, Z_2, Z_3, Z_4 automatically passes through Z_5. For $i \neq j$ write L_{ij} for the line joining Z_i, Z_j. Then the line-pair comprising the lines L_{13}, L_{24} passes through Z_5, so either Z_1, Z_3, Z_5 are collinear, or Z_2, Z_4, Z_5 are collinear. Either way we contradict the Component Lemma, that a line meets an irreducible conic in at most two points. $\qquad\square$

Example 17.1 The zero sets of the standard ellipses, standard hyhperbolas, and standard parabolas of Examples 4.2, 4.3, 4.1 were all shown to be infinite. Thus their equations are determined up to scalar multiples.

Exercises

17.1.1 Find the conic which passes through the points $(2, 3)$, $(3, 2)$, $(3, 1)$, $(1, 3)$, $(1, 2)$ and verify that it passes through $(2, 1)$.

17.1.2 Clearly, the hypothesis of the Uniqueness Theorem can be weakened, by assuming only that the zero set contains five (not necessarily distinct) points, no three of which are collinear. Show that it suffices to assume a yet weaker hypothesis, that the zero set contains at least five (not necessarily distinct) points, no four of which are collinear.

17.2 Proof of Invariance

Our second proof is that of the Invariance Theorem, that the three invariants τ, δ, Δ do not change when a congruence is applied to a quadratic polynomial.

Theorem 17.2 *Let Q, Q' be strictly congruent quadratic functions with invariants τ, δ, Δ and τ', δ', Δ'. Then $\tau = \tau'$, $\delta = \delta'$, $\Delta = \Delta'$. (The Invariance Theorem.)*

The mechanics of the proof are simplified by using matrix notation, both for quadratic functions and for congruences. First, we can write the general quadratic function ($\star$) in a concise matrix form. Recall that in Chapter 4 we associated to Q the 3×3 symmetric matrix

$$A = \begin{pmatrix} a & h & g \\ h & b & f \\ g & f & c \end{pmatrix}.$$

Now write z for the row vector $z = (x, y, 1)$, and z^T for its transposed column vector. The reader will readily check then that Q can be written

$$Q(x, y) = zAz^T. \tag{17.2}$$

Furthermore, we can write the general congruence (14.7) in the following matrix form

$$z^T = PZ^T, \tag{17.3}$$

where $Z = (X, Y, 1)$ and P is the matrix of coefficients

$$P = \begin{pmatrix} \cos\theta & \sin\theta & u \\ -\sin\theta & \cos\theta & v \\ 0 & 0 & 1 \end{pmatrix}. \tag{17.4}$$

Note that the matrix P has the property that its determinant $\det P = 1$. We use this fact in the following proof of the Uniqueness Theorem.

Proof To clarify its structure, we split the proof into three steps, one for each of the three invariants.

Step 1   We use the matrix form (17.2) for the quadratic function Q, and likewise (17.3) for a general congruence. Applying the congruence to Q, we obtain the quadratic polynomial

$$Q'(X, Y) = zAz^T = Z(P^T A P)Z^T.$$

Thus $A' = P^T A P$ is the matrix associated to Q'. Using the multiplicative property of the determinant, and $\det P = \det P^T = 1$, we have

$$\Delta' = \det A' = \det(P^T A P) = \det P^T \det A \det P = \det A = \Delta.$$

Step 2   Now let B, B' be the leading 2×2 submatrices of A, A', and let S be the leading 2×2 submatrix of P. Then, by inspection, $B' = S^T B S$. Using the multiplicative property of the determinant again, and $\det S = \det S^T = 1$, we have

$$\delta' = \det A' = \det(S^T B S) = \det S^T \det B \det S = \det B = \delta.$$

Step 3   It remains to prove that $\tau' = \tau$. Recall from linear algebra that the trace $\tau(U)$ of a 2×2 matrix U is the sum of its diagonal entries. We use the fact that for two such matrices U, V we have $\tau(UV) = \tau(VU)$. Then, using the notation of Step 2, and the fact that SS^T is the identity matrix

$$\tau' = \tau(B') = \tau(S^T B S) = \tau(SS^T B) = \tau(B) = \tau.$$

$\square$

The reader who goes further down the geometry road will discover that there is rather more to the invariants τ, δ, Δ than is expressed by the Invariance Theorem. There is a real sense in which they are the *only* invariants of conics. More precisely, any polynomial expression in the coefficients of a general conic Q with the same 'invariance' property can be expressed solely in terms of τ, δ, Δ.

Index